T0134091

Introduction to Non-Invasive EEG-Based Brain–Computer Interfaces for Assistive Technologies

Introduction to Non-Invasive EEG-Based Brain–Computer Interfaces for Assistive Technologies

Edited by
Teodiano Freire Bastos-Filho

CRC Press
Taylor & Francis Group
Boca Raton London New York

CRC Press is an imprint of the
Taylor & Francis Group, an **informa** business

First edition published 2021
by CRC Press
6000 Broken Sound Parkway NW, Suite 300, Boca Raton, FL 33487-2742

and by CRC Press
2 Park Square, Milton Park, Abingdon, Oxon, OX14 4RN

© 2021 Taylor & Francis Group, LLC

CRC Press is an imprint of Taylor & Francis Group, LLC

Library of Congress Cataloging-in-Publication Data
Names: Bastos-Filho, Teodiano Freire, editor.
Title: Introduction to non-invasive EEG-based brain-computer interfaces for assistive technologies / edited by Teodiano Freire Bastos-Filho.
Description: Boca Raton : CRC Press, 2020. | Includes bibliographical references and index. | Summary: "This book provides an overview about different applications of brain-computer interfaces (BCIs) based on the authors more than 20 years of research in the field"—Provided by publisher.
Identifiers: LCCN 2020009393 (print) | LCCN 2020009394 (ebook) | ISBN 9780367501976 (hardback) | ISBN 9781003049159 (ebook)
Subjects: MESH: Brain-Computer Interfaces | Electroencephalography—methods| Self-Help Devices
Classification: LCC RC386.6.E43 (print) | LCC RC386.6.E43 (ebook) | NLM WL 26.5 | DDC 616.8/047547—dc23
LC record available at https://lccn.loc.gov/2020009393
LC ebook record available at https://lccn.loc.gov/2020009394

ISBN: 978-0-367-50197-6 (hbk)
ISBN: 978-0-367-50222-5 (pbk)
ISBN: 978-1-003-04915-9 (ebk)

Typeset in Times
by codeMantra

Contents

Preface

This book aims to bring to the reader an overview about different applications of brain–computer interfaces (BCIs) based on our experience of more than 20 years working on these interfaces. First, in Chapter 1, a review of the human brain and EEG (electroencephalogram) signals is conducted, describing the human brain, anatomically and physiologically, with the objective of showing some of patterns of EEG signals used to control BCIs. The chapter introduces aspects such as planes of section and reference points of the human brain; primary somatosensory cortex (S1), primary motor cortex (M1), primary auditory cortex (A1), primary visual cortex (V1), and the areas of Wernicke and Broca; contralaterality of motor movements (pyramidal tract and contralaterality of motor movements); neuronal circuits and oscillatory activity of the thalamo-cortical system; circuits that are involved in motor activity (direct pathway of movement); EEG; EEG electrodes; EEG acquisition; main EEG rhythms; artifacts present in the EEG signals; methods for EEG filtering, such as spatial filters; signal-to-noise ratio (SNR); event-related potential (ERP); movement-related (cortical) potential (MRP/MRCP); event-related desynchronization/synchronization (ERD/ERS); and steady-state visual evoked potential (SSVEP) and its types: dependent SSVEP and independent SSVEP.

Following, in Chapter 2, the BCI is introduced, and in Chapter 3 different applications of BCIs are discussed, which were developed in our laboratory, such as a BCI based on ERD/ERS patterns in α rhythms, used to command a robotic wheelchair and an augmentative and alternative communication (AAC) system onboard the wheelchair; a BCI based on dependent SSVEP to command the same robotic wheelchair; a BCI based on SSVEP to command a telepresence robot and its onboard AAC system; a BCI based on SSVEP to command an autonomous car; a BCI based on independent SSVEP, using depth-of-field (DoF), to command the same robotic wheelchair of our laboratory; the use of compressive technique in our SSVEP-based BCI; a BCI based on motor imagery (using different techniques) to command a robotic monocycle and a robotic exoskeleton; and our first steps to build a neurorehabilitation system based on motor imagery of pedaling together an immersive virtual environment with avatar. This book is concluded with a wide discussion, in Chapter 4, about the future of the non-invasive BCIs.

I believe this book can help the readers have new insights on other applications for BCIs.

Teodiano Freire Bastos-Filho
Universidade Federal do Espírito Santo (UFES), Brazil

Acknowledgments

I would like to thank UFES/Brazil for giving me a leave of 3 months, allowing me to write this book. I would also like to thank all the collaborators of this book. Without their supports, this book would not have been possible.

Editor

Teodiano Freire Bastos-Filho received his degree in electrical engineering from the Universidade Federal do Espírito Santo (UFES), Vitória, Brazil, in 1987, and his Ph.D. in physical sciences from the Universidad Complutense de Madrid, Madrid, Spain, in 1994. He works in the Department of Electrical Engineering, Postgraduate Program in Electrical Engineering and Postgraduate Program in Biotechnology at UFES, Vitória, Brazil, and he is a research productivity fellowship 1-D in the Brazilian National Council for Scientific and Technological Development (CNPq). His research interests include signal processing, rehabilitation robotics, and assistive technologies.

Contributors

Teodiano Freire Bastos-Filho
Electrical Engineering Department
Postgraduate Program in Electrical
  Engineering
Postgraduate Program in
  Biotechnology
Universidade Federal do Espírito
  Santo
Vitória, Espírito Santo, Brazil

Alexandre Luís Cardoso Bissoli
Postgraduate Program in Electrical
  Engineering
Universidade Federal do Espírito Santo
Vitória, Espírito Santo, Brazil
and
National Institute of Industrial Property
Rio de Janeiro, Brazil

Thomaz Rodrigues Botelho
Postgraduate Program in Electrical
  Engineering
Universidade Federal do Espírito Santo
Vitória, Espírito Santo, Brazil
and
Instituto Federal do Espírito Santo
São Mateus, Espírito Santo, Brazil

Alessandro Botti Benevides
Postgraduate Program in Electrical
  Engineering
Universidade Federal do Espírito Santo
Vitória, Espírito Santo, Brazil
and
Universidade Federal do Pampa
Alegrete, Rio Grande do Sul, Brazil

Leandro Bueno
Postgraduate Program in Electrical
  Engineering
Universidade Federal do Espírito Santo
and
Instituto Federal do Espírito Santo
Vitória, Espírito Santo, Brazil

Eliete Caldeira
Electrical Engineering Department
Universidade Federal do Espírito Santo
Vitória, Espírito Santo, Brazil

Vivianne Cardoso
Postgraduate Program in
  Biotechnology
Universidade Federal do Espírito Santo
Vitória, Espírito Santo, Brazil

Javier Ferney Castillo Garcia
Postgraduate Program in Electrical
  Engineering
Universidade Federal do Espírito
  Santo
Vitória, Espírito Santo, Brazil
and
Universidad Santiago de Cali
Cali, Colombia

Anibal Cotrina Atencio
Postgraduate Program in Electrical
  Engineering
Universidade Federal do Espírito Santo
Vitória, Espírito Santo, Brazil
and
Universidade Federal do Espírito Santo
São Mateus, Espírito Santo, Brazil

**Eduardo Henrique Couto
Montenegro**
Postgraduate Program in Electrical
 Engineering
Universidade Federal do Espírito
 Santo
Vitória, Espírito Santo, Brazil
and
BA Systèmes
Mordelles, France

Celso De La Cruz Casaño
Postgraduate Program in Electrical
 Engineering
Universidade Federal do Espírito
 Santo
Vitória, Espírito Santo, Brazil
and
Pontificia Universidad Católica del Perú
Huancayo, Peru

Denis Delisle Rodríguez
Postgraduate Program in Electrical
 Engineering
Universidade Federal do Espírito
 Santo
Vitória, Espírito Santo, Brazil

Maria Dolores Pinheiro de Souza
Postgraduate Program in Psychology
Universidade Federal do Espírito Santo
Vitória, Espírito Santo, Brazil
and
Espaço Trate Trauma
Vitória, Espírito Santo, Brazil

André Ferreira
Electrical Engineering Department
Universidade Federal do Espírito
 Santo
Vitória, Espírito Santo, Brazil

Alan Silva da Paz Floriano
Postgraduate Program in Electrical
 Engineering
Universidade Federal do Espírito
 Santo
Vitória, Espírito Santo, Brazil
and
Instituto Federal do Espírito Santo
São Mateus, Espírito Santo, Brazil

Anselmo Frizera-Neto
Electrical Engineering Department
Postgraduate Program in Electrical
 Engineering
Universidade Federal do Espírito
 Santo
Vitória, Espírito Santo, Brazil

Christiane Mara Goulart
Postgraduate Program in
 Biotechnology
Universidade Federal do Espírito Santo
Vitória, Espírito Santo, Brazil
and
Faculdade Multivix
Vila Velha, Espírito Santo, Brazil

Dharmendra Gurve
Department of Electrical, Computer &
 Biomedical Engineering
Ryerson University
Toronto, Ontario, Canada

Muhammad Asraful Hasan
Department of Electrical, Computer &
 Biomedical Engineering
Ryerson University
Toronto, Ontario, Canada
and
University of Adelaide
Adelaide, South Australia, Australia

Kevin Antonio Hernández-Ossa
Postgraduate Program in Electrical
    Engineering
Universidade Federal do Espírito Santo
Vitória, Espírito Santo, Brazil
and
Virtual Reality Rehab, Inc.
Clermont, Florida, United States of
    America

Sridhar Krishnan
Department of Electrical, Computer &
    Biomedical Engineering
Ryerson University
Toronto, Ontario, Canada

Jéssica Paola Souza Lima
Postgraduate Program in
    Biotechnology
Universidade Federal do Espírito Santo
Vitória, Espírito Santo, Brazil

Berthil Longo
Postgraduate Program in Biotechnology
Universidade Federal do Espírito Santo
and
Secretaria de Estado da Educação
Vitória, Espírito Santo, Brazil

Flávia Aparecida Loterio
Postgraduate Program in
    Biotechnology
Universidade Federal do Espírito Santo
Vitória, Espírito Santo, Brazil

Sandra Mara Torres Müller
Postgraduate Program in Electrical
    Engineering
Universidade Federal do Espírito Santo
and
the Instituto Federal do Espírito Santo
Vitória, Espírito Santo, Brazil

Jeevan Pant
Department of Electrical, Computer &
    Biomedical Engineering
Ryerson University
Toronto, Ontario, Canada

Alexandre Geraldo Pomer-Escher
Postgraduate Program in Biotechnology
Universidade Federal do Espírito
    Santo
and
Panpharma Distribuidora de
    Medicamentos
Vitória, Espírito Santo, Brazil

**Richard Junior Manuel
Godinez-Tello**
Postgraduate Program in Electrical
    Engineering
Universidade Federal do Espírito
    Santo
Vitória, Espírito Santo, Brazil
and
Instituto Federal do Espírito Santo
Serra, Espírito Santo, Brazil

Hamilton Rivera-Flor
Postgraduate Program in Electrical
    Engineering
Universidade Federal do Espírito Santo
Vitória, Espírito Santo, Brazil

Alejandra Romero-Laiseca
Postgraduate Program in Electrical
    Engineering
Universidade Federal do Espírito
    Santo
Vitória, Espírito Santo, Brazil
and
Virtual Reality Rehab, Inc.
Clermont, Florida, United States of
    America

Mario Sarcinelli-Filho
Electrical Engineering Department
Postgraduate Program in Electrical
  Engineering
Universidade Federal do
  Espírito Santo
Vitória, Espírito Santo, Brazil

Ana Cecilia Villa-Parra
Postgraduate Program in Electrical
  Engineering
Universidade Federal do Espírito Santo
Vitória, Espírito Santo, Brazil
and
Universidad Politécnica Salesiana
Cuenca, Azuay, Ecuador

1 Review of the Human Brain and EEG Signals

Alessandro Botti Benevides, Alan Silva da Paz Floriano, Mario Sarcinelli-Filho, and Teodiano Freire Bastos-Filho

CONTENTS

The purpose of this chapter is to describe the human brain, anatomically and physiologically, to better understand some of the patterns observed in EEG (electroencephalogram) signals used to distinguish distinct mental tasks to control a BCI (brain–computer interface).

In our body, the nervous system, comprising the central nervous system (CNS) and the peripheral nervous system (PNS), coordinates and monitors all our conscious and unconscious activity. The CNS consists of the brain and spinal cord, and the PNS consists of nerves[1] and ganglia[2] [1]. In the next section, we will focus on CNS and its main constituent, the brain.

---

[1] Groupings of axons in the PNS. Only one group of CNS axons is named as nerve, which is the optic nerve [1].

[2] Greek word meaning "node", which is an agglomerate of cell bodies of neurons found outside the CNS [1].

1.1   PLANES OF SECTION AND REFERENCE POINTS OF THE HUMAN BRAIN

The relative position of brain structures is located through planes of section and reference points. This section covers some terms used throughout this book to locate and describe brain structures. The portion of the brain facing forward, regarding the human body, is called anterior, and the portion facing backward is called posterior. On the other hand, the direction facing upward is called dorsal and the direction facing downward is called ventral. Figure 1.1 shows the directions and the three planes of section that are the sagittal, coronal, and horizontal. The structures nearest to the medial line are called medial structures, and structures furthest from the medial line are called lateral structures. The structures that are on the same side of the medial line are called ipsilateral to each other, and the structures that are on opposite sides of the medial line are contralateral to each other. Finally, similar structures that are on both sides of the medial line are bilateral.

The brain surface is composed of numerous circumvolutions, which are the evolutionary result of the brain's attempt to increase its cortical area, being confined to the skull. The protrusions are called gyri, and the grooves are called sulci; very deep sulci are called fissures. The exact pattern of gyri and sulci may vary considerably from individual to individual, but many features are common to all human brains.

By convention, the brain is divided into lobes, based on the overlying skull bones: the central sulcus separates the frontal lobe from the parietal[3] lobe; and the lateral sulcus, or Sylvian fissure,[4] separates the frontal lobe and the temporal[5] lobe; and the occipital[6] lobe is located on the caudal region of the brain, and is surrounded by the parietal and temporal lobes [1].

The temporal lobe receives and processes auditory information, which is related to object identification and naming. The frontal lobe (including the motor, the premotor, and prefrontal cortexes) is involved in planning actions and movements, as well as abstract thought. The parietal lobe is the primary somatosensory cortex and receives information about touch and pressure from thalamus, and the occipital lobe receives and processes visual information [2].

The cerebral surface or cortex[7] is organized like a patchwork quilt, which were first identified and numbered by Brodmann[8] (Figure 1.2). The main areas related to the processing of the senses are the primary motor cortex, or M1 (area 4), the supplementary motor area (SMA), and premotor area (PMA) (area 6) in the frontal lobe; the primary somatosensory cortex, or S1 (areas 1, 2, and 3), and the primary gustatory

---

[3] The term "parietal" is derived from Latin, "parietalis", meaning wall.

[4] Assigned in tribute to Franciscus Sylvius (1614–1672), who was a Dutch physician and scientist.

[5] The term "temporal" arises from Latin "tempus" meaning time. The word "time" was used for this region because it is typically on the sides of the skull where hair first becomes gray, showing the ravages of time.

[6] The term "occipital" means something situated near the "occiput", which is derived from Latin prefix "ob" combined with "caput" meaning "at the back of head".

[7] The term "cortex" is derived from Latin, meaning "bark" [1].

[8] Korbinian Brodmann (1868–1918) was a German neurologist and psychiatrist responsible for the subdivision of the cerebral cortex in 47 functional areas, called Brodmann's areas, which were numbered according to the sequence in which he studied them [1].

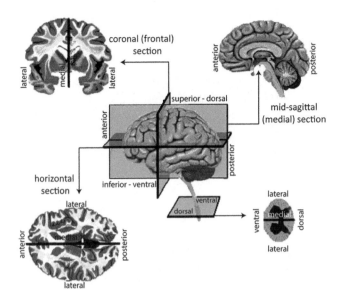

FIGURE 1.1 Planes of section and reference points of the human brain.

cortex (area 43) in the parietal lobe; the primary auditory cortex, or A1 (areas 41 and 42), and the olfactory cortex in the temporal lobe; and the primary visual cortex, or V1 (area 17), in the occipital lobe.

In the context of the mental tasks addressed in this book, the areas 4 (M1) and 6 (SMA and PMA) are related to motor mental tasks, whereas areas 39 and 40 (Wernicke's area), and 44 and 45 (Broca's area) are related to the tasks of imagination of words, which are detailed in the next section. On the other hand, the areas 41 and 42 (A1) are related to music imagery tasks, and the area 17 (V1) is related to visual tasks (Figures 1.2 and 1.3).

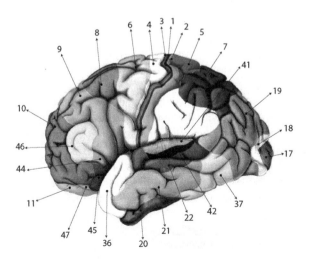

FIGURE 1.2 Brodmann's cytoarchitectonic map. (Adapted from [1].)

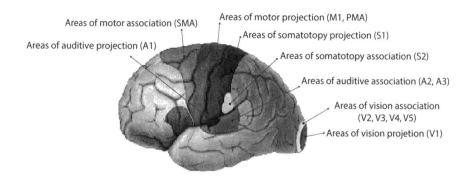

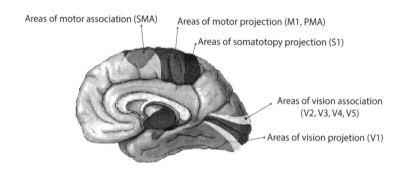

FIGURE 1.3   Main areas related to the processing of the senses. (Adapted from [1].)

1.2   DETAILS OF S1, M1, A1, V1, WERNICKE'S, AND BROCA'S AREAS

The primary motor cortex is directly responsible for the coordination of voluntary movements. The left side of Figure 1.4 shows the somatotopic[9] map of M1, which correlates some M1 areas with the control of body parts. It is worth noting that more than a half of M1 comprises the control of muscles linked to hands and speech [2].

The right side of Figure 1.4 shows the somatotopic map of S1, correlating areas of the somatosensory cortex with the sensitivity of various areas of the body. Notice that the somatotopic organization of human precentral gyrus (M1) is very similar to that observed in somatosensory areas of postcentral gyrus (S1). Electrophysiological changes of areas corresponding to movements of hands and feet are located in the precentral gyrus, since the same area in the postcentral gyrus corresponds to the sensitivity of touch, pressure, and temperature of such limbs.

---

[9] The mapping of body surface sensations or control of body movement in a CNS area is called somatotopy [1].

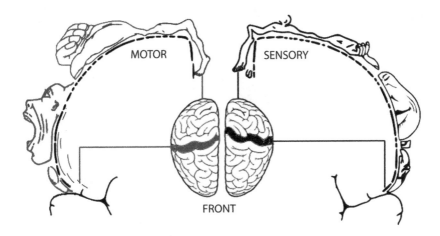

FIGURE 1.4 Left: Somatotopic map of human precentral gyrus (M1); right: somatotopic map of human postcentral gyrus (S1).

The PMA, or premotor cortex, has the function of supporting the movements generated by the primary motor cortex of both hemispheres, making possible the execution of a "motor imagery" task, which is a "simulation" of the muscular movement to be performed. The signals associated with this "motor imagery" task are directly conveyed from PMA to M1 to excite multiple muscle groups related to accomplishment of the task [1].

Human studies performed by the Danish neurologist Per Roland using positron emission tomography (PET) to track changes in cortical activation patterns that follow voluntary movements showed that performing finger movements increases blood flow in the following regions: somatosensory areas, posterior parietal cortex (PPC); portions of prefrontal cortex; and the areas M1, SMA, and PMA [3]. When participants were asked to just mentally imagine the movement without actually moving the fingers, the area of SMA and PMA remained active, while the area of M1 did not remain active [1].

The language processing, comprehension, and speech production occur in Broca's area, whereas the association and interpretation of information occur in Wernicke's area (Figure 1.5a), which plays a very important role during the chaining of the discourse. This area allows us to understand what others say and also provides the ability to organize the words in a way syntactically correct. Broca's area is located in the left hemisphere in 95% of persons [2].

The primary auditory cortex (A1) is the first cortical region of the auditory pathway, which is directly connected with the medial geniculate nucleus (MGN) of the thalamus. It roughly corresponds with Brodmann's areas (Figure 1.2) and is located on the temporal lobe. This cortex area is the neural crux of hearing and, in humans, language, and music. The right auditory cortex has long been shown to be more

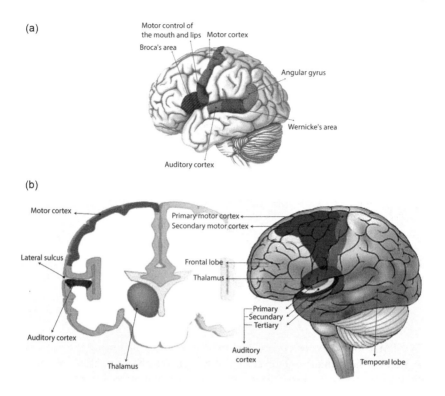

FIGURE 1.5 (a) Location of Broca's area and Wernicke's area. (Adapted from [1].) (b) Primary, secondary, and tertiary auditory cortexes.

sensitive to tonality, whereas the left auditory cortex has been shown to be more sensitive to minute sequential differences in sound, such as in speech. The auditory cortex is divided into three separate parts: primary, secondary, and tertiary auditory cortexes. These structures are formed concentrically around one another, with the primary cortex in the middle and the tertiary cortex on the outside (Figure 1.5b).

The primary auditory cortex is tonotopically[10] organized, which means that neighboring cells in the cortex respond to neighboring frequencies, forming a "frequency map". This brain area is thought to identify the fundamental elements of music, such as pitch and loudness. The secondary auditory cortex (A2) has been indicated in the processing of harmonic, melodic, and rhythmic patterns. The tertiary auditory cortex (A3) supposedly integrates everything into the overall experience of music and remembering a sound stimulus, and only faintly activates the tertiary auditory cortex.

The visual information processing begins with the sensitizing of photosensitive cells in retina (rods and cones), sending information through the optic nerve to the

---

[10] Tonotopy is the spatial arrangement of where sounds of different frequencies are processed in the brain. Tones close to each other in terms of frequency are represented in topologically neighboring regions in the brain. Tonotopic maps are a particular case of topographic organization, similar to somatotopy in somatosensory areas and retinotopy in the visual system.

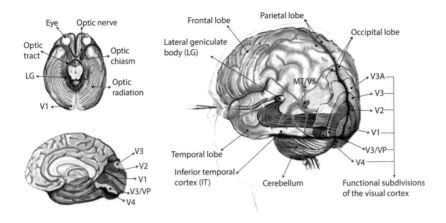

FIGURE 1.6 Visual information path from retina to visual cortex.

thalamus area called lateral geniculate nucleus (LG) or body.[11] The LG then modulates and transmits visual information to the striate cortex or primary visual cortex (V1). Visual areas are subdivided into V1, V2, V3, V4, and V5, also known as visual area MT (middle temporal) (see Figure 1.6). The V1 projections extend to V2, V3, and V5 areas. V4 is related to the perception of shape and color, and the inferior temporal (IT) cortex is related to visual memory and recognition of human faces.

1.3 PYRAMIDAL TRACT AND CONTRALATERALITY OF MOTOR MOVEMENTS

This section discusses the contralaterality of motor movements, which means that the motor area in the right cerebral hemisphere controls the voluntary movements of the left side of the body; conversely, the motor area in the left cerebral hemisphere controls the voluntary movements of the right side of the body. Thereby, the imagination of movements of the right hand is processed in the primary motor cortex of the left hemisphere. It is worth commenting that while the motor cortex of each hemisphere activates movements of the opposite side, the PPC activates movements of both sides of the body [15,16].

The somatic motor system (SMS) and vegetative nervous system (VNS) are all neural references of CNS. The brain sends signals for muscle control and receives sensory information through 12 pairs of cranial nerves and 31 pairs of nerves in the spinal cord (Figure 1.7a). Axons that carry stimulatory signals from the brain to effector organs, such as muscles and glands, through the spinal cord, are primary efferent[12] nerves of the SMS. Primary nerves enter the spinal cord through ventral roots. Axons that carry information from the sensory receptors of the skin,

---

[11] The lateral geniculate nuclei are formed by six layers of overlapping cells, which curve around the optic tract, as the articulation of a knee. From this fact derives the name geniculate, "geniculatus" from Latin, meaning "like a little knee" [1].

[12] The term "efferent" is derived from Latin, meaning "which brings" [1].

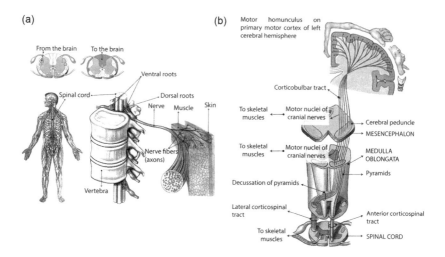

FIGURE 1.7   (a) Spinal nerves and spinal nerve roots. (Adapted from [1].) (b) Pyramidal system.

muscles, and joints to the brain, through the spinal cord, are the primary afferent[13] nerves of the somatic sensory system, which enter the spinal cord through dorsal roots. Thus, the two roots of the spinal cord transmit information in opposite directions [1].

The path involving the transmission from the motor cortex to the muscles is called corticospinal tract[14] or pyramidal tract. The transmission occurs directly in the pyramidal tract and indirectly through multiple accessory pathways involving basal ganglia, cerebellum, and several brainstem nuclei. The pyramidal tract originates from pyramidal neurons in layer V of the cerebral cortex of M1 (30%), PMA and SMA (30%), and S1 (40%). The pyramidal tract is mainly composed of motor axons, constituting the volunteer component of the motricity. The pyramidal pathways consist of a single tract, originating in the brain, which is divided into two separate tracts in the spinal cord: the lateral corticospinal tract and the anterior corticospinal tract (Figure 1.7b).

From all fibers of the pyramidal tract, 80% cross its side in the decussation[15] of pyramids in the Bulb[16] (contralaterally), forming the lateral corticospinal tract, and 20% follow caudally to lateral funiculus of the spinal cord (ipsilaterally), forming the anterior corticospinal tract. The anterior corticospinal tract also crosses its side, but only at medullar level, where it makes synapse [2]. Therefore, the behavior of the pyramidal pathways leads to the conclusion that the voluntary motricity is 100% crossed, either at the bulbar level or at the spinal cord level.

---

[13] The term "afferent" is derived from Latin, meaning "which leads" [1].

[14] Grouping of CNS axons that have common origin and destination [1].

[15] Axons crossing from one side to the other [1].

[16] The cross section of the medulla at the decussation level shows that the corticospinal tract forms a triangular protuberance, which is why this area is called bulbar pyramid and the corticospinal tract is called the pyramidal tract [1].

1.4 NEURONAL CIRCUITS AND OSCILLATORY ACTIVITY OF THE THALAMOCORTICAL SYSTEM

Although some research groups have used patterns from the PPC (where the movements start to be planned) [15,16], some online BCIs use the paradigm of motor imagery recognition, taken into account the event-related desynchronization or event-related synchronization (ERD/ERS) patterns. This section discusses the physiological aspects that cause these patterns, observed during the imagination of motor tasks. Subsequently, after the presentation of the characteristics of the EEG signal, the ERD/ERS patterns will be defined and mathematically quantified.

Neurons that connect the nervous system and different layers of the brain form the neuronal circuits, which transmit information through excitatory and inhibitory synapses. Excitatory synapses may be mediated by the neurotransmitter acetylcholine (ACo), dopamine (DA), noradrenaline (NA), adrenaline, serotonin, glutamate (Glu), and glycine (Gly), whereas the inhibitory synapses are mediated by the neurotransmitter gamma-aminobutyric acid (GABA).

In some neuronal circuits, the input signal causes an excitatory synapse in one direction and an inhibitory synapse in another direction (Figure 1.8a). In this figure, the input fiber (sensory neuron) directly excites the neuron #1 and simultaneously excites the intermediate inhibitory neuron (neuron #2), which secretes GABA to inhibit neuron #3. This kind of circuit is important to prevent excessive activity in many parts of the brain [2].

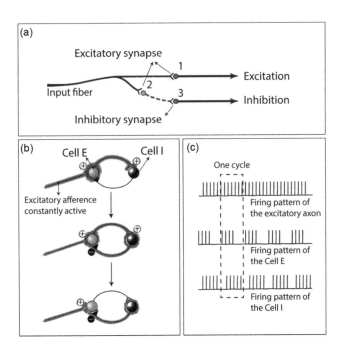

FIGURE 1.8 (a) Excitatory–inhibitory neuronal circuit [2]; (b) two neurons oscillator; (c) firing pattern of the two neurons oscillator. (Adapted from [1].)

Oscillatory neuronal circuits are the most important neuronal circuits of the nervous system. Figure 1.8b shows a very simple circuit, consisting only of an excitatory neuron, an inhibitory neuron, and an external constant afference, where an excitatory cell (cell E) and an inhibitory cell (cell I) establish synapses with each other. As long as there is a constant excitatory conduction over the cell, which does not have to be rhythmic, the resulting activity of the set tends to oscillate. A cycle of activity across this network generates the firing pattern shown in Figure 1.8c.

Oscillatory neuronal circuits can also be formed by positive feedback. Consequently, once stimulated, the circuit can produce periodic stimuli for long periods. The simplest oscillatory circuit is shown in Figure 1.9a. This circuit only has one neuron, whose part of output extends to its own dendrites to be re-stimulated. Figure 1.9b shows a circuit with additional feedback neurons, Figure 1.9c shows a little more complex circuit with facilitators and inhibitors neurons, and Figure 1.9d shows an oscillatory circuit with multiple parallel neurons [2].

Kevan Hashemi calculated that coherent neurons could not activate themselves at a frequency much higher than 100 Hz. The activation of a single neuron takes roughly 2 ms for its rising and falling edges, whereas its refractory period, in which it cannot be reactivated, is about 10 ms. Then, it will prevent any neuron from firing much faster than 100 Hz [4]. Simple neuronal circuits working with negative (Figure 1.9b) or positive (Figure 1.9a) feedback will also obey to the same upper bound frequency.

The oscillatory activity can be measured in the EEG signal, being an emergent property of the thalamocortical system and the corticocortical system. In the thalamocortical system, the oscillatory neuronal circuits are formed between the thalamus and cortex. On the other hand, in the corticocortical system, the oscillatory neuronal circuits are formed between the different layers of the cerebral cortex.

The human cortex is a laminar structure composed of six distinct layers[17] of alternate white and gray laminae with different kinds of neurons. Figure 1.10 shows a

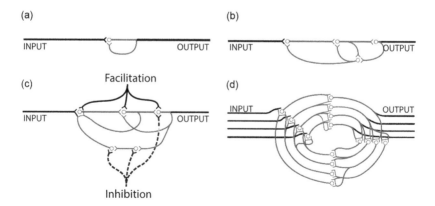

FIGURE 1.9   Oscillatory neuronal circuits. (Adapted from [2].)

[17] The cortex layers were discovered in 1840 by the French physician Jules Baillarger. His name is associated with the inner and outer bands of Baillarger, which are two layers of white fibers of the cerebral cortex. They are prominent in the sensory cortical areas because of high densities of thalamocortical fiber terminations.

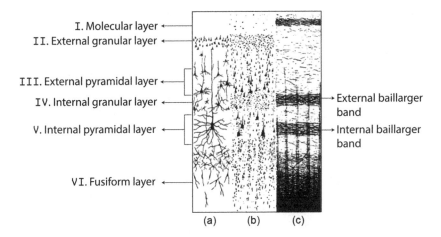

I. Molecular layer
II. External granular layer
III. External pyramidal layer
IV. Internal granular layer
V. Internal pyramidal layer
VI. Fusiform layer

External baillarger band
Internal baillarger band

(a) (b) (c)

FIGURE 1.10 Cortical layers. (a) Golgi's method; (b) Nissl's method; (c) Weigert's method. (Adapted from [5].)

schematic representation of the cortical layers, as it appears in histological preparations stained by different methods [5]. Perpendicular to the layers there are large neurons called pyramidal neurons, which connect the various layers together, representing about 85% of the neurons in the cortex [2]. Corticocortical oscillatory circuits typically do not involve a large number of cells working in synchrony; thus, the set does not induce high electrical amplitude activity and is hardly measured by electrodes on the scalp, which has little contribution to the EEG.

Under certain conditions, thalamic neurons can generate precisely rhythmic discharges of action potentials (APs) that reach the cortex. Thalamus is located in the center of the brain, connecting its different parts, and all information reaching the cortex passes through it. Information from the sensory systems is conveyed to the thalamus, which redirects it to specific areas of the cerebral cortex, and information about the control of voluntary movement traverses the thalamus in the opposite direction.

The thalamus is mainly composed of gray matter,[18] in which multiple cores are distinguished. Figure 1.11a shows the location of the thalamus in the brain. Figure 1.11b shows the main nuclei of the thalamus, and Figure 1.11c shows the connectivity of each nucleus with the cortex. The information from M1 passes through the ventral lateral (VL) nucleus of the thalamus, where it is directed to some cranial nerves (III, IV, V, VII, IX, X, XI, and XII) and the spinal cord [1].

To illustrate the process of information flow through thalamus, a quick mention of the senses of tasting, hearing, and vestibular and somatosensory systems will be given. Three cranial nerves (VII, IX, and X) carry the gustatory information of different regions of the tongue and oral cavity to the bulb after rising to the ventral posteromedial (VPM) nucleus of the thalamus and finally reaching the primary gustatory cortex (Figure 1.11c). Auditory information captured by auditory receptors

---

[18] The gray matter consists of cell bodies of neurons, whereas the white matter is formed by myelinated axons.

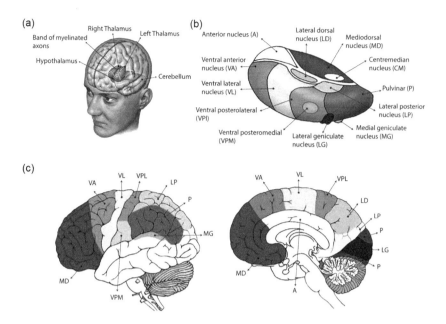

FIGURE 1.11 (a) Location of the thalamus in the brain; (b) schematic representation of the main nuclei of the thalamus; (c) connectivity of each nucleus with the cortex.

of the cochlea[19] is driven by the auditory nerve (cranial nerve VIII) for the MGN of the thalamus, from where it goes to the primary auditory cortex (A1) (Figure 1.11c).

The vestibular system, which informs the position and movement of the head providing sense of balance, is driven by the auditory nerve to the ventral posterolateral (VPL) and VPM nucleus of the thalamus, projecting axons to regions close to the representation of the face between S1 and M1. The somatic sensory system, which involves the senses of touch, temperature, pain, and body position, sends the information from the sensory receptors through the spinal cord, following to medulla, pons, and midbrain until it reaches VPL and VPM nuclei, projecting axons to S1 [1].

The thalamus has a particular set of neuronal cells called pacemaker neurons. These circuits operate as the circuits of Figures 1.8 and 1.9, providing self-sustaining rhythmic discharge, even in the absence of external afferent. The rhythmic activity of these pacemaker circuits becomes synchronized with many other thalamic cells that are projected to the cortex by thalamocortical axons. Thus, a relatively small group of centered thalamic cells can compel a much larger group of spread out cortical cells to follow the thalamic pace [1]. When a neuron or a neuron mass starts to spike at a fixed delay in response to periodic input, as the thalamic pace, it is called phase-locking.

The large number of cortical cells working in synchrony induces a high electrical amplitude activity that can be measured by electrodes on the scalp. This synchrony represents a large contribution to the EEG signal and is closely related to ERD/ERS patterns observed during the performance of mental tasks. Hereinafter, for a better

---

[19] The term "cochlea" is derived from Latin, meaning "snail" [1].

understanding of what occurs during mental tasks, an example of a pattern observed in the EEG due to thalamocortical rhythms during the opening and closing of eyes will be mentioned.

The visual information captured in the retina is carried through the optic nerve (cranial nerve II) and then by optical tracts to the LG nucleus of the thalamus, which projects axons to layers IV and VI of V1. Thus, when the eyes are open, the nerve impulses are continually transmitted to the visual cortex. While activated, the neuronal circuits block the sending of rhythmic activity from thalamus to V1, which undoes the synchrony.

Local brain activity increases greatly because the neural masses of V1 are processing a lot of information, but the synchronization between each neuron becomes so small that the resulting induced electrical activity measured in the scalp is almost null. The results are EEG waves of small-amplitude, high-frequency, and irregular rhythm known as β (14–30 Hz) rhythm.

When the eyes are closed, no impulses are being transmitted to the visual cortex. Then, the neural circuits allow sending rhythmic activity from thalamus to V1, which become synchronized, or phase-locked, in the same frequency band of the thalamic pace. Local brain activity decreases, but the neuronal circuits of V1 are synchronized at nearby frequencies, resulting in a high-energy pace, the α (8–12 Hz) rhythm.

The resulting pattern indicates that when the eyes are open, the EEG signal measured over V1 shows small amplitude at the frequencies of the α rhythm, since the neural masses are desynchronized. When the eyes are closed, the EEG signal will measure large amplitude at the frequencies of the α rhythm, since the masses are synchronized with the thalamic pace. The same association between high information processing and low energy at the frequencies of α rhythm can also be seen over M1, SMA, and PMA during the imagination of motor tasks, being the physiological basis for the phenomenon known as ERD/ERS. In the next section, the motor circuit will be analyzed in the same way as the visual circuit to understand how the ERD/ERS pattern occurs during motor mental tasks.

1.5 DIRECT PATHWAY OF MOVEMENT

This section analyzes the circuits that are involved in motor activity, linking different areas of the motor cortex. Actually, the initiation of voluntary movements engages areas in frontal, prefrontal, and parietal cortices, which are connected to the basal nuclei,[20] deep in the brain. The basal nuclei receive most of its input signals from the cerebral cortex and return almost all of their output signals to the cerebral cortex. In each hemisphere, the basal nuclei are formed by the caudate nucleus, putamen,[21] globus pallidus,[22] subthalamic nucleus, and substantia nigra,[23] which are located around the thalamus, occupying a large portion of the internal regions of both brain hemispheres (Figure 1.12a). The caudate and putamen together are called the striatum, which is the target from the cortical afference to the basal nuclei [2].

---

[20] Improperly called basal ganglia, as the term "ganglia" is only used for neural cluster in the PNS [5].
[21] The term "putamen" is derived from Latin "putare", meaning "to prune, to think, or to consider".
[22] The term "globus pallidus" is derived from Latin, meaning "pale globe".
[23] The term "substantia nigra" is derived from Latin, meaning "black substance" [1].

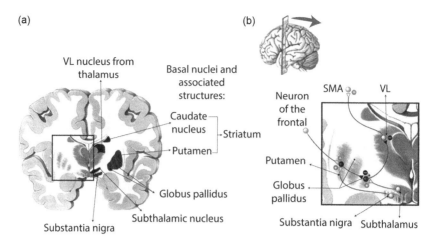

FIGURE 1.12 (a, b) Motor circuit. Synapses marked with sign (+) are excitatory, whereas synapses marked with sign (−) are inhibitory. (Adapted from [1].)

The basal nuclei project an afference, called VLo, to the VL nucleus of the thalamus. The VLo afference is a projection of axons from the basal nuclei through VL until SMA, which is intimately interconnected with M1 (Figure 1.12b). Thus, there is a path on which information will cycle from the cortex to the basal nuclei and the thalamus, which will come back to cortex, such as follows [1]:

$$Cortex\left(frontal, prefrontal, and\ parietal\right) \rightarrow Striate \rightarrow Globus\ pallidus \rightarrow$$

$$VLo \rightarrow Cortex\left(SMA\right)$$

This neuronal loop is known as the direct pathway of movement. The direct pathway originates with an excitatory connection to cortical cells in the putamen, and the cells of the putamen establish inhibitory synapses on neurons in the globus pallidus, which in turn makes inhibitory connections with the cells of the VL. The thalamocortical connection from VL to the SMA is excitatory, facilitating the trigger of cells related to movements in the SMA (Figure 1.12a). The functional consequence of cortical activation of the putamen is the excitement of the SMA by VL. This is because neurons in globus pallidus at rest are spontaneously active and thus inhibit VL.

The cortical activation (frontal, prefrontal, or parietal) excites neurons of the putamen, which inhibit neurons of the globus pallidus, removing the inhibition of VL and allowing VL neurons to become active. The activity in VLo drives the activity to SMA. Thus, this part of the circuit acts as a positive feedback loop that can serve to focus the activation of spread cortical areas to the SMA. It is speculated that the signal to start the motor activity occurs when activation of the SMA is driven above some threshold, for the activity that reaches it through the basal nuclei [1].

There is a direct afference from thalamus to M1, which is mainly originated by another part of VL, called VLc, which retransmits the information from cerebellum.

The information that comes from the cerebellum is related to motor learning and ballistic movements, which are executed so fast that the feedback cannot act to control the movement. For these movements, the cerebellum relies on predictions based on experience, comparing what was intended with what happened, conducting to the learning.

The cerebellum has one-tenth of the total volume of the brain, but it has a high density of neurons in the cortex (in fact more than 50% of the total number of neurons of the CNS). To clarify the importance of the cerebellum, the path that connects the cerebellum to other parts of the brain has 20 times more axons than the pyramidal tract [1]. The path that connects the cerebellum to M1 forms another important motor circuit; however, as the movements controlled by the cerebellum are not related to motor imagination, the analysis of the direct pathway of movement is limited here.

In brief, during the performance of a motor imagery task, the initial cognitive activity that manifests the intention of performing the task is originated in the frontal cortex. This activity propagates to the putamen through excitatory connections. The activation of the putamen inhibits the globus pallidus, which is spontaneously active. Then, the inactivation of the globus pallidus makes it no longer inhibit VLo, which stays active and can retransmit the signals to specific areas of SMA. Thus, the neural masses SMA will be working with different aspects of a complex cognitive task, and the neurons will be firing quickly, but not simultaneously, which results in a low synchrony.

Conversely, when no motor mental task is being performed, there is no transmission of information from the frontal, prefrontal, or parietal cortex to the putamen. And when the putamen is inactive, it does not inhibit the globus pallidus, but inhibits VLo. The inactivity of VLo allows SMA to receive signals from the thalamic pacemaker neuronal circuits, as SMA is connected to M1 through the afferences that extend into layer V of M1. Therefore, the influence of the thalamic pacemaker signal can be observed in pyramidal neurons of layer V of M1, which will present a synchronous behavior.

The large number of pyramidal neurons in layer V of SMA, PMA, and M1 working in synchrony induces a high electrical amplitude activity that can be measured by electrodes on the scalp. Then, when no motor mental task is being performed, the EEG signal measured over SMA, PMA, and M1 will show high amplitude at the frequencies that are in synchrony. This frequency interval, about 8–12 Hz, was measured empirically and is known as μ rhythm [6]. On the other hand, during the performance of a mental task of motor imagery, the neural masses of SMA, PMA, and M1 will be desynchronized and will induce a low electrical amplitude activity over this area. Then, the EEG will have low amplitude for the μ rhythm (8–13 Hz).

Although the μ rhythm lies almost on the same frequency interval of the aforementioned α rhythm, there is a distinction between them, which is the location at which they are measured, as the μ rhythm is observed on the sensorimotor cortex and the α rhythm is observed on the visual cortex, but also almost on the entire scalp. Despite the different nomenclatures for this frequency interval, both of them are due to the projection of the thalamic pace on the cortex. Section 1.9 will give more details about EEG rhythms.

Finally, this section aims to clarify the physiologic basis behind clear patterns that occurs during motor mental tasks and can be measured by the EEG. That is, during the motor mental task, the μ rhythm will present low amplitude relative to the rest state, in which no specific mental task is being performed.

1.6 EEG SIGNAL

The EEG or electroencephalography is the recording of the electrical activity of a large population of neurons (several million neurons) of the cerebral cortex measured on the surface of the scalp using electrodes. This noninvasive technique is most usual; however, much more accurate neuronal activity can be obtained by introducing the electrode within the brain tissue (depth recording) or by placing electrodes on the exposed surface of the brain, which is called electrocorticogram (ECoG) [7]. These implantable electrodes can be formed by microelectrode array (e.g., 4×4 mm) with about 100 electrodes of 1.5 mm in length, able to register between 100 and 200 neurons [14,15].

Although implantable electrodes provide quite accurate information about neuron activities, they require to be surgically placed on the user's scalp, have a limited period of use, and need to be recalibrated for each use [14,15]. On the other hand, the EEG acquisition is a noninvasive relatively simple technique: about two or more dozen electrodes are easily placed in standard positions on the scalp, and voltage fluctuations, usually a few tens of microvolts (μV), are measured between selected pairs of electrodes and then amplified dozen thousand times [8]. A typical EEG recording (Figure 1.13) is a set of many irregular simultaneous tracings, indicating changes in voltage between pairs of electrodes. Each output signal of the amplifier controls a record pen or is stored in a computer memory [1].

The EEG mainly records the extracellular currents that arise as a consequence of synaptic activity in dendrites of neurons in the cerebral cortex. The extracellular electric field is mainly generated by the postsynaptic potential (PSP) that may be excitatory postsynaptic potential (EPSP) or inhibitory postsynaptic potential (IPSP). When the AP (100 mV) reaches postsynaptic dendrites, it causes a current flow that enters through the synapse to the postsynaptic dendrites of the next neuron. This current is called

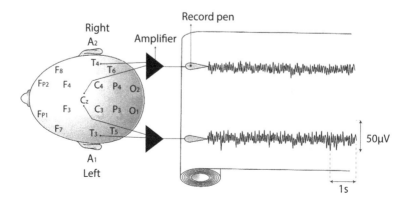

FIGURE 1.13   Typical EEG recording. (Adapted from [1].)

extracellular activation current, which is of around 3 nA, and causes the rising potential within the postsynaptic membrane, that is, the EPSP. The current then enters the postsynaptic dendrites toward the soma[24] and goes out into the extracellular fluid by the membrane capacitance of the cell, and then returns to the dendrites, making a circular path. This circular current is called excitatory postsynaptic current [4] (Figure 1.14).

The neuron has a rest potential (about 65 mV), but it does not produce an external electrical field. The cell membrane is formed by a 5 nm phospholipid bilayer closely approximated by the infinite parallel plate capacitor with a capacitance of around 0.01 F/m² between the intracellular and extracellular fluid. Thus, the electric field outside the capacitor is zero, and the rest potential has no influence on the EEG signal. In the same way, during the activation for the propagation of an AP, the interior of the cell jumps up by 100 mV, but it does not induce an electric field outside the cell membrane [4].

As mentioned earlier, the neurons of the cortex are divided into six layers parallel to the surface. These neurons are of two types: pyramidal (layers III and V) and non-pyramidal (layers I, II, IV, and VI). The glial cells have spherical symmetry, not having a common direction for the propagation of the electrical signal. As a consequence of this symmetry, the resulting electromagnetic field is null [9]. On the other hand, the PSP of the neurons generates extracellular electric fields that have a bipolar distribution; thus, neurons can be modeled like a small dipole, whose orientation depends on the type of PSP that is occurring in the neuron [10].

For example, EPSP occurs in synapses mediated by the neurotransmitter glutamate. Glutamate activates the opening of cation channels, allowing the flow of Na+ ions into the postsynaptic dendrite. The flow of Na+ makes the soma positive in relation to extracellular fluid (Figure 1.15a). IPSP occurs in GABA-mediated synapses, allowing the flow of Cl− ions into the postsynaptic dendrites, and the flow

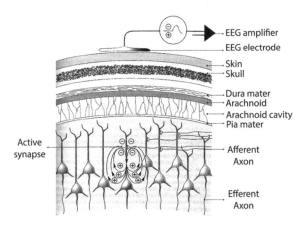

FIGURE 1.14 Electric field generated by extracellular currents in pyramidal cells. (Adapted from [1].)

---

[24] The word "soma" comes from the Greek language, meaning "body". The soma of a neuron is often called the cell body.

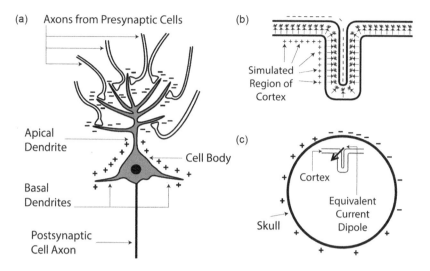

FIGURE 1.15 (a) Bipolar configuration of the electric field of pyramidal neuron during a PSP; (b) bipolar configuration of the electric field of a pyramidal neurons of an active cortical region; (c) equivalent current dipole of the active cortical region. (Adapted from [10].)

of Cl− makes the soma negative in relation to extracellular fluid [1]. Therefore, the orientation of dipoles of excitatory and inhibitory neurons is opposite [10].

The dipole of an individual neuron is impossible to be measured by electrodes on the scalp; however, under specific conditions, the dipoles of many neurons are added, generating a resulting field that can be measured on the scalp. For that, the dipoles need to be spatially aligned and the PSP should occur at approximately the same time, so that the dipole activity adds up. If the neurons are randomly oriented, the positivity of a dipole would be canceled by the negativity of the adjacent dipole. Furthermore, if a neuron is stimulated by an excitatory neurotransmitter, and an adjacent neuron is stimulated by an inhibitory neurotransmitter, the dipoles will have opposite orientations and cancel each other. However, if many neurons having similar orientations and the same type of neurotransmitter are stimulated at approximately the same moment, then the dipoles will be added and their activity may be measured on the scalp [10].

Nonpyramidal cells are mostly inhibitory, mediated by the neurotransmitter GABA, with the exception of interneurons that can be excitatory or inhibitory. These cells have bipolar distribution and, in majority, use a common neurotransmitter; however, their random orientation cancels the electromagnetic field resulting from the dipoles.

Pyramidal cells are excitatory and use a common neurotransmitter, the glutamate. They have a resultant electric field with bipolar configuration (Figure 1.15a) and are spatially aligned perpendicularly in the cortex (Figure 1.10). Thus, they are the main contributors to the formation of the electrical signals recorded as the EEG. Most of pyramidal cells have their axons directed to the thalamus and basal nuclei, so that the soma relatively positive is below the postsynaptic dendrites that are relatively negative and closer to the surface of the cortex. Figure 1.15b shows the bipolar configuration of the electric field of the pyramidal neurons in an active cortical region, in which the outer surface is negative and the inner region is positive. Figure 1.15c shows the equivalent dipole resulting from this region [10].

Extracellular electric fields generated by neurons are attenuated and scattered when crossing the skull toward the scalp, due to the low conductivity of the skull that acts as a low-pass spatial filter [8]. The skull thickness can vary from 3.3 to 6 mm, depending on its location, causing a variation of electrical resistance [11]. Therefore, the scalp distribution of the electric field is mainly distorted by variations in the conductivity of the skull, but it is also affected by the conductivity of the meninges[25] and the skin (Figure 1.14).

1.7 EEG ELECTRODES

The typical EEG electrode is made of a very conductive material to measure the electrical potential difference induced by the residual electric field of the scalp, relative to a reference point. However, only with the contribution of small voltages of thousands of cells firing together, this signal can be sufficiently intense to be detected on the surface of the scalp. This population of neurons is called the neural mass and consists of 10^4 to 10^7 neurons [12,42], and the joint activation of neurons in a neural mass is called synchronism. If the synchronous excitation of this group of cells is repeated several times, the EEG will be consisted of large rhythmic waves [1].

Figure 1.16 shows the generation of large EEG signals by neural synchronous activity. Figure 1.16a shows six pyramidal neurons wherein the PSP is measured

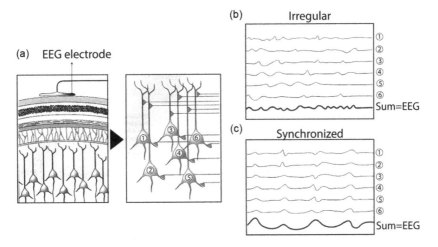

FIGURE 1.16   (a) Pyramidal neurons beneath EEG electrode; (b) unsynchronized neuronal activity; (c) synchronized neuronal activity. (Adapted from [1].)

---

[25] Meninges is the plural of meninx, from Greek "membrane". The meninges are the system of membranes that envelop and protect the CNS (spinal cord, brainstem, and brain), consisting of three layers: the dura mater, the arachnoid mater, and the pia mater.

between a pair of sensors at their extremities (gray triangles). In Figure 1.16b, the neurons are activated in irregular time intervals and the summed PSP activity of all six neurons has small amplitude. In Figure 1.16c, the neurons are activated synchronously; thus, the summed PSP activity of all six neurons has high amplitude [1].

The silver electrodes wrapped in silver chloride (Ag/AgCl) were used until the 1980s, since silver is the best conductor of electricity among the metals, with a resistivity (ρ) of 1.59×10^{-8} Ωm. For a comparison, the white matter resistivity is about 6.5 Ωm, the cortex (gray matter) resistivity is about 3.0 Ωm, the cerebrospinal fluid resistivity is about 0.64 Ωm and the skull resistivity is about 120 Ωm [4].

The internal impedance comprises the impedance of the white matter, gray matter, cerebrospinal fluid, meninges, and skull, whereas the external impedance comprises the contact impedance and the impedance of electrodes, wires, and the amplifier system. The model of the external impedance is a resistance only, because the effect of the brain's electrical permittivity and permeability is negligible at EEG frequencies, and so, a capacitor and an inductor are not included in the equivalent circuit.

For EEG signal acquisition, the gold standard uses electrodes fixed to the scalp through a gel or a conductive electrolytic paste, which are intended to reduce the contact impedance/resistance between the electrode and the skin. Usually, the value of this impedance must be lower than 5 kΩ, to be considered a good resistance matching [10,13]. As the amplifiers of the EEG acquisition circuit have very high impedance (typically higher than 100 kΩ), variations in impedance of the electrodes in the order of a few thousand ohms have negligible effect on the measured voltage [8].

Due to the laborious time spent to acquire EEG signals through gelled electrodes, in 2009, Emotiv Systems launched a neuroheadset (Epoch) with wet electrodes (saline-soaked), which became quite popular. However, this headset only allows short-term experiments, as the saline starts to dry after 15–20 min (compared to several hours of electrodes with gel). On the other hand, recently, EEG dry electrodes have arisen in the market, which make unnecessary the use of gel, paste, or saline, however, at a very expensive cost: a cap with 16 dry electrodes and accessories for EEG acquisition can cost some thousands of US dollars.

1.8   EEG ACQUISITION

To acquire EEG signals, the first step was given in 1958 by Herbert Jasper, who suggested a system for naming and placement of electrodes on the scalp, which is now worldwide used, called "International 10–20 System" [17]. Figure 1.17a shows that electrodes on the edges of the scalp are 10% distant of the horizontal line connecting the nasion to the inion through the preauricular point, where this percentage is related to the length of the line connecting the nasion to the inion through the vertex. All electrodes are positioned at a distance of 20% between each other. Then, numbers "10" and "20" of the "International 10–20 System" refer to these percentage values. The electrodes are named by a capital letter corresponding to the initial of the brain lobe where they are placed (F = Frontal, C = Central, P = Parietal, O = Occipital, and T = Temporal), followed by an even number for the right hemisphere and an odd number for the left hemisphere. Electrodes on the frontal pole are named Fp, and the letter "A" is used for electrodes placed in the ear

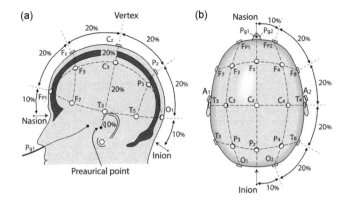

FIGURE 1.17   (a, b) International 10–20 System.

(from "auricular"). For the electrodes in the line connecting the nasion to the inion, the letter "z" is added, which indicates "zero", rather than a number, indicating the central division of the brain hemispheres (Figure 1.17b) [10].

If the electrical potential of the scalp was measured with only one EEG electrode relative to the ground of the acquisition circuit, the circuit would measure only the static electricity difference between the scalp and the circuit, which is much larger than the neural activity. For comparison, the neural activity generates voltage changes in the order of tens of microvolts, whereas the average static electricity of a human body is around 4 to 35 thousand volts.[26]

However, even using the voltage difference between two electrodes in the scalp to create an EEG channel, any noise that affects the ground or the power of the acquisition circuit would also mask the neural activity. To solve this problem, the EEG amplifier system uses differential amplifiers with three electrodes (Figure 1.18) to create a channel: an electrode called "active"[27] electrode (A), one reference electrode (R), and a ground electrode (G). Thus, the differential amplifier amplifies the voltage difference between V_{AG} and V_{RG} ($C = V_{AG} - V_{RG}$), and the common noise that affects the ground of both measures is eliminated [10].

The ground electrode is usually positioned on the frontal bone to minimize the noise with muscular origin, and its potential is canceled during differential amplification. Then, its location is not as important as the location of the reference electrode. There is no ideally "neutral" place to position the reference electrode, so it should be

---

[26] The average electrical resistance of a human body is between 1.3 and 3 kΩ, and the average capacitance is around a few hundreds of pF. Then, the capacitance model of the human body, as defined by the Electrostatic Discharge Association (ESDA), is a 100 pF capacitor in series with a resistor having a resistance of 1.5 kΩ charged with voltages from 4 to 35 kV. In this model, a capacitor with an initial voltage of 35 kV provides an electrical current below 5 mA after 1.3 μs. Thus, the high voltage acquired by static electricity from the human body provides currents above the threshold of human perception, which is around 5 mA, during instants of time lower than 1.3 μs, and no electrical discharge is noticed.

[27] To distinguish itself from the ground and reference electrodes, even measuring passively the scalp potential, this electrode is called "active". It should not be confused with active electrodes, which perform the signal amplification within itself.

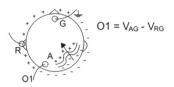

$O1 = V_{AG} - V_{RG}$

FIGURE 1.18 Grounding electrode (G), the reference electrode (R), and the active electrode (A) used to compose the occipital channel O1.

taken in mind that the EEG signal from one channel always reflects the contribution of both active and reference electrodes. The reference electrode is usually placed on an ear lobe or both. Figure 1.18 shows the grounding electrode (G), the reference electrode (R), and the active electrode (A) on the scalp area of interest to measure the EEG. The ground electrode is positioned on the frontal region, the reference electrode is positioned on the left ear, and the electrode (A) is positioned on the occipital lobe of the left hemisphere to create the channel O1. This figure also shows the resulting dipole of an active area of the cortex and its potential distribution on the scalp. In this example, V_{AG} is lower than V_{RG}, and therefore, the calculated potential for O1 will be negative.

The EEG signal is represented by a number sequence ordered with respect to its temporal evolution. Actually, the potential $V_{AG}(t)$ and $V_{RG}(t)$ and the channel O1(t) are shown along the text without reference to its time instant just for simplicity. Then, in this book, all operations involving variables without the time reference are considered non-recursive, being applied only to the same time instant.

For the EEG acquisition with multiple locations, there are three distinct ways of electrode derivations to create the channels: bipolar method, unipolar method (or common electrode/reference), and free reference methods, which apply spatial filter, such as Laplacian, local average reference (LAR), and common average reference (CAR) (Figure 1.19), among others. Figure 1.19a illustrates the bipolar method in which each channel (T3, C3, C4, and T4) is the difference in potential between two neighboring electrodes (A1, A2, A3, A4, or A5). It may be noted that in this method, there is no fixed reference electrode, and the number of channels is always smaller than that of electrodes. It must be remembered that the grounding electrode is present for all three kinds of derivation.

Figure 1.19b shows the unipolar method, which has a reference electrode (R) common to all channels, and the measured voltage is the difference between any electrode and the reference. Figure 1.19c shows the unipolar method with biauricular reference electrodes (R1 and R2), in which the measured voltage is the difference between any electrode and the average values of the reference electrodes. In the CAR method, the signal is originally acquired by the unipolar method (Figure 1.19b, channels C3, Cz, and C4), and then channels C3′, Cz′, and C4′ are formed by subtracting the unipolar potentials by the average reference potential (M) (Figure 1.19d). In general, considering n channels,

$$C_i' = C_i - \frac{1}{n}\sum_{j=1}^{n} C_i \qquad (1.1)$$

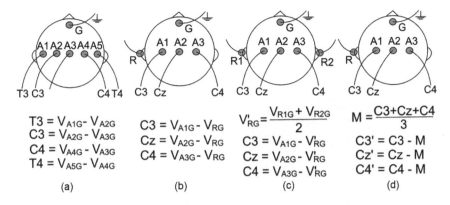

FIGURE 1.19 (a) Bipolar method; (b) unipolar method (uniauricular reference); (c) unipolar method (biauricular reference); (d) CAR method using the uniauricular reference.

where C_i is the ith channel, and C_i' represents the ith channel using the CAR. As the common average prioritizes signals that are present in a large number of channels, the subtraction leads to the elimination of these signals, working as a spatial filter; that is, it accentuates components with highly localized distributions while eliminating components that are in most channels, as indeed the biological external noise [18]. Thus, a time-varying external noise, which equally affects all electrodes, is eliminated by the CAR method. Suppose that there is a noise X that affects the grounding potential of the acquisition system, V_{XG}, and a biological noise component B from the human body that equally affects the potential of all electrodes on the scalp, V_{BG}, then the potential of the electrodes (A) and reference (R) will be

$$\begin{cases} V_{AG} = \hat{V}_{AG} + V_{BG} + V_{XG} \\ V_{RG} = \hat{V}_{RG} + V_{BG} + V_{XG} \end{cases} \tag{1.2}$$

where $\hat{V}$ represents the potential measured by the electrode in the absence of noise. Thus, the channel C' calculated with the CAR method will be given by

$$C_i' = C_i - \frac{1}{n} z \sum_{i=1}^{n} C_i = V_{AG} - V_{RG} \left(\frac{1}{n} \sum_{j=1}^{n} V_{iG} - V_{RG} \right) \leftrightarrow$$

$$C_i' = V_{AG} - \frac{1}{n} \sum_{i-1}^{n} V_{iG} \leftrightarrow$$

$$C' = \hat{V}_{AG} + V_{BG} + V_{XG} - \left(\frac{1}{n} \sum_{j=1}^{n} \hat{V}_{iG} + V_{BG} + V_{XG} \right) \leftrightarrow$$

$$C' = \hat{V}_{AG} - \left(\frac{1}{n} \sum_{j=1}^{n} \right) \rightarrow$$

$$C' \approx \hat{V}_{AG} \tag{1.3}$$

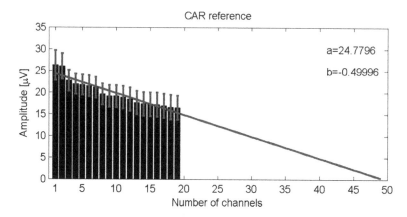

FIGURE 1.20   Decrease of the average reference amplitude as the number of channels increases.

In (1.3), the channel C' will be free of noise X affecting the grounding electrode, and also the biological noise B that ideally equally affects all electrodes. Finally, it is assumed that the activity of different brain areas is uncorrelated and can be considered random, then, as the number of electrodes increases, the averaged reference (M) approaches zero. Then, the sum

$$\left(\frac{1}{n} \sum_{j=1}^{n} \right)$$

can be discarded and the potential measured in channel C' increasingly becomes independent of the fluctuations of the reference.

Figure 1.20 shows the decrease of the mean reference (M) amplitude as the number of channels increases. It can be seen that M tends to a constant other than zero. This is because the ideal assumptions that the activity of the measures of the brain areas is uncorrelated and fully random and the biological external noise equally affects all electrodes are not completely valid. This figure was obtained with the EEG of 60 repetitions of motor mental tasks related to the right-hand imagination of movements. This figure shows that the reference amplitude decreases from 26.3 to 16.5 µV when using 19 EEG channels. The error bars are the standard deviations calculated from that experiment.

A linear regression was performed, such as shown in Figure 1.20, in which the intercept (a) and the slope (b) are approximately 24.8 and −0.5, respectively. This linear regression gives estimation in which, for each added channel, the reference amplitude should decrease 0.5 µV. Although those ideal assumptions cannot be assured, according to Figure 1.20, with 50 roughly independent channels, the reference amplitude should be close to zero or at least smaller than the current one.

1.9   MAIN EEG RHYTHMS

The EEG signal has specific features in defined frequency bands. It is known that the performance of some activities, such as sleep, relaxation, or mental effort, is related to specific frequency bands, which can even be induced, blocked, or changed during a mental task [6]. The main frequency bands are denoted by Greek letters: α, β, γ, δ, μ, and θ. Each band is generally observed at a specific location and circumstance; thus, δ rhythm is observed on the frontal lobe during deep sleep; α rhythm is observed mainly on the occipital lobe when the eyes are closed, and μ rhythm is observed on the motor cortex during the performance of a motor activity. The normal oscillatory activity (Figure 1.21) is classified into infra-slow (0.02–0.1 Hz), slow (0.1–15 Hz), fast (20–60 Hz), and ultra-fast (100–600 Hz). Spindles are bursts of neural oscillatory activity that are generated by interplay of the thalamic reticular nucleus (TRN) and other thalamic nuclei during stage 2 NREM (non-rapid eye movement) sleep in a frequency range of approximately 11–16 Hz. On the other hand, ripples are oscillatory patterns in the brain hippocampus during immobility and sleep, in a frequency range of approximately 140–200 Hz.

The most important EEG features are comprehended in frequencies that extend to 30 Hz. This frequency range is subdivided into groups or rhythms (δ, θ, α, μ, and β) related to the location of measurement and the frequency band, such as shown in Table 1.1.

Regarding the ultrafast activity (>100 Hz), Kevan Hashemi [4] reports that the highest frequency fluctuation observed in EEG goes up to 120 Hz. He showed that the high-frequency oscillations (HFO) and very high-frequency oscillations (VHFO) reported to exist in animal and human EEG by some authors (e.g. see [20]) may be artifacts of band-pass filtering or a mismeasurement, i.e., an artifact of electromyographic activity related to other neural function, such as minute eye movements.[28]

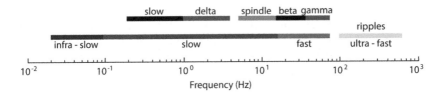

FIGURE 1.21   Coverage of the EEG frequency spectrum. (Adapted from [19].)

TABLE 1.1

Frequencies Occurring in the Human Brain

Band	Frequency (Hz)	Amplitude (μV)	Localization
Delta (δ)	1–4	<100	Variable
Theta (θ)	4–8	<100	Variable
Alpha (α)	8–12	20–60	Occipital
Beta (β)	14–30	20–30	Frontal and parietal
Mu (μ)	8–13	<50	Cortex motor
Gamma (γ)	25–100	<50	Variable

---

[28] Minute eye movements, also known as saccade, are fast eye movements of a few minutes of arc, in which there is no conscious control of the speed.

The activation period of a single neuron is about 2 ms; then, the activation frequency of a neuron cannot be higher than 500 Hz. The refractory period of a single neuron is of order 10 ms, which will prevent any neuron from firing much faster than 100 Hz. Thus, as EEG is generated by large numbers of neurons acting coherently, EEG should not include a fundamental oscillation frequency much higher than 100 Hz [20]. Therefore, HFO and VHFO of EEG will not be addressed in this book. Following, the mental states associated with the δ, θ, α, μ, and β rhythms are described.

δ **Rhythm:** frequency band between 1 and 4 Hz registered in individuals in deep sleep and also may appear related to some pathological states.

θ **Rhythm:** frequency band between 4 and 8 Hz, whose signal has higher amplitude than δ and β rhythms. It is found on the frontal region during mental activities, such as problem-solving and in the temporal and parietal regions during emotions of stress, disappointment, and frustration (see [67,68], for example, for the evaluation of θ rhythm to infer stress).

α **Rhythm:** the International Federation of Electroencephalography and Clinical Neurophysiology defines the α rhythm as the frequency band between 8 and 12 Hz, occurring in awake people on posterior regions of the brain, with typically higher voltage over the occipital areas. The amplitude is variable, but it is almost always below 50 μV in adults. It is more easily detected with eyes closed and with subject in conditions of physical relaxation and mental inactivity. It is blocked or attenuated by attention, especially visual, and by mental effort [2]. Also, the mental imagination of sound generally elicits an increase of α rhythm activity [21].

μ **Rhythm:** frequency band between 8 and 13 Hz. It is a rhythm associated with motor activities and best acquired in the motor cortex. It is blocked or attenuated with movement or the intention to move. As mentioned in Section 1.5, although its frequency range and amplitude are almost similar to α rhythm, the μ rhythm is topographically and physiologically different from α rhythm [18].

β **Rhythm:** frequency band between 14 and 30 Hz, with lower amplitude, and usually caused by the opening of the eyes, being in a state of wakefulness, or REM (rapid eye movement) sleep. It is blocked by motor activity and tactile stimulation [2].

γ **Rhythm:** frequency band between 25 and 100 Hz. The low γ band (25–40 Hz) has been shown to be present during the perception of sensory events and the process of recognition.

Figure 1.22 shows these brain rhythms and the attenuation of α rhythm while the eyes are open. The replacement of the α rhythm for an asynchronous β rhythm of small amplitude when the eyes are open is discussed in Section 1.4.

In general, the rhythms of low amplitude and high frequency are associated with wakefulness and alertness or the stage of sleep in which dreams occur. This is because when the cortex is more actively involved in the processing of information, may these be brought by sensory afferents, or generated by internal processes, and the level of activity of cortical neurons is relatively high, but also desynchronized.

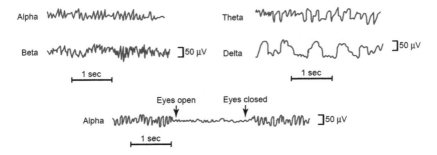

FIGURE 1.22 Brain rhythms α, β, θ, and δ. The EEG signal at the bottom shows the replacement of the α rhythm by the β rhythm. (Adapted from [2].)

Each neuron or a very small group of neurons is heavily involved in a slightly different aspect of a complex cognitive task, firing rapidly, but not simultaneously, in relation to its neighboring neurons. This leads to a low synchrony, and thus, the amplitude of the EEG is low and the β rhythm prevails [1].

In contrast, rhythms of high amplitude and low frequency are associated with stages of dreamless sleep and the pathological state of coma. Thus, during deep sleep, cortical neurons are not busy processing information, and most of them are excited phasic by a slow and rhythmic afference. This rhythmic signal or pacemaker is originated in the oscillatory neuronal circuits of thalamus, which imposes itself on the neurons of the cortex. Then, the synchrony is high, and therefore, the amplitude of the EEG is also high [1].

1.10 ARTIFACTS

The EEG signal is easily corrupted by other electrical signals, due to its small amplitude (in the order of μV). The noise found on EEG is called "artifacts", which belong to two categories based on their source: physiological (also called biological or internal) or nonphysiological (also called technical or external) [43, 22, 23]. Technical artifacts occur due to external electrical interference or malfunction of the EEG device (electrodes, wires, amplifiers, filters, power supply), with grid line interference and fluctuations in electrode impedance the most important. Grid line artifact is caused by magnetic interference from sources of AC voltage. This artifact shows a typical frequency of 50 or 60 Hz, depending on the frequency of the grid line. It can be removed by the use of properly tuned filters without significant loss in the EEG signal information [28]. Alternatively, short wires can be used between the electrode and the amplifier, or the measurements can be performed in a shielded room. Figure 1.23 shows the EEG channels Fp1 and C3 in which the grid line artifact is present in the 60 Hz component. These EEG signals were acquired during the mental task of imagination of movement of the right hand (between 5 and 25 s), using the unipolar method with the reference electrode on the left ear lobe and the ground electrode on the user forehead.

Artifacts due to fluctuations in electrode impedance are usually caused by poor fixation of electrodes and by sweating, in which the junction's skin–electrolyte and electrolyte–electrode cause a DC level in the electrode. This DC level reaches values

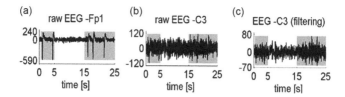

FIGURE 1.23 Eye blink artifacts on EEG. (a) EEG of channel Fp1 heavily contaminated with low frequencies from eye blink artifacts; (b) EEG of channel C3 contaminated with low-frequencies eye blink artifacts and artifact from the power grid line; (c) EEG of channel C3 after applying the CAR filter and high-pass filter with a cutoff frequency of 5 Hz.

from 0.1 to 1.7 V, which is much higher than the EEG amplitude; therefore, it is necessary to use high-pass filters at the input of the amplifiers [9]. Traditionally, the EEG systems configure their high-pass input filters by adjusting the time constant (τ) between 0.32 s −0.5 Hz, 0.16 s −1.0 Hz, and 0.08 s −2.0 Hz.[29]

On the other hand, the physiological artifacts are originated by interference of the EEG with other biological signals, such as cardiac, pulse, respiratory, sweat, glosso-kinetic, eye movement (blink, lateral rectus spikes from lateral eye movement), and muscle and movement [22]. Figure 1.23 shows the presence of grid line artifacts and spikes related to eye movements in the EEG signals. Five eye blink artifacts can be seen in the raw EEG of channel Fp1 (which have low-frequency components, usually below 10 Hz). These artifacts represent the major peaks observed in the EEG signal. It is important to mention that the signals from both channels (Fp1 and C3) were collected simultaneously in the same experiment. The channel C3 does not present visible eye blink artifacts, and then, the frontal channel Fp1, that is closer to the artifact source, was much more affected. In addition to eye movements, muscle artifacts and electrocardiogram (ECG) are the other main physiological artifacts, which are detailed as follows.

Artifacts from eye movements: there is a continuous potential difference between cornea and retina that forms an electrical dipole for each eye. The cornea-retinal potential (CRP) is between 10 and 30 mV, which is much higher than the EEG amplitudes. The rotation of the eye causes a rotation of the dipole, generating a signal known as electrooculogram (EOG), which propagates through the skin to closer electrodes, causing an increase or decrease in the EEG baseline [24].

Muscle artifacts: these are electrical signals related to muscle contraction (acquired through surface electromyography – sEMG), particularly due to movements of the head, neck, or eye blinks. For this reason, during EEG acquisition, the individual is asked not to move or blink their eyes. On the other hand, it is known that electrodes at temporal (T3, T4, T5, and T6) and lateral-frontal sites (F7 and F8) are particularly affected by facial muscle artifacts, from tension or jaw movements [8]. In general, frontal and temporal electrodes are more heavily contaminated by artifacts than central scalp channels [25].

ECG artifacts: the cardiac electrical activity can also be recorded on the scalp, as it affects the EEG; another ECG-related artifact is the ballistocardiogram (BCG).

---

[29] The first-order high-pass filter cutoff frequency (f_c) is given by: $f_c = 1/2\pi\tau$, $\tau = RC$, where R is the resistance and C is the capacitance.

Both artifacts are due to heart pulsations, which generate micro-movements that affect the EEG electrodes and wires.

Beyond the interference of technical and physiological artifacts, the EEG is also affected by the electrical activity of the brain itself. The EEG of an area of interest is a mixture of unrelated signals from cortical neighboring areas that are scattered around and attenuated by the skull and scalp. This is considered a special type of artifact, in which there are no exact solutions for unmixing it from the EEG signal. This problem is known as the inverse problem that traditionally has infinite solutions, due to the nature of its variables. A particular inverse solution uses the calculation of local field potential (LFP), which is invasive recording of the electric potential in the extracellular space in brain tissue [26]. Another solution uses the distribution of cortical extracellular currents, known as cortical current density (CCD) [27].

1.11   SPATIAL FILTERING

Filtering methods are aimed to eliminate the frequency components in which the artifacts are present. For this purpose, high-pass, low-pass, band-pass, or band-reject filters are used. As mentioned earlier, technical artifact as the interference from the grid line has a well-defined spectral component around 50 or 60 Hz (depending on the country). Thus, a notch filter can be used to eliminate this kind of artifact. On the other hand, physiological artifacts, such as ECG, sEMG, and EOG, have overlapping spectral components over the EEG spectrum. Figure 1.24 shows an analysis of coherence between the EEG signal measured by electrodes F3 and C3, using the bipolar method, as well as the sEMG signal measured bilaterally on the deltoid, while an individual was performing extension movements of the fingers [28]. It may be noted that the spectra overlap throughout the range of frequencies (Figure 1.24b) and the higher coherence is obtained between 6 and 16 Hz, completely corrupting the α and μ rhythms [29]. Thus, although the filtering techniques are not able to remove all

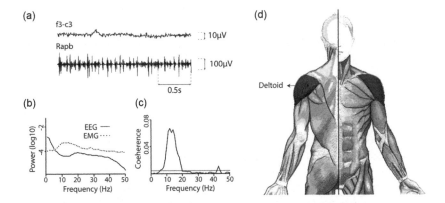

FIGURE 1.24   (a) EEG (from F3 to C3 electrodes) and sEMG (from right abductor pollicis brevis – Rapb) signals; (b) EEG and sEMG spectra; (c) coherence between EEG and sEMG. The horizontal line is the 95% confidence interval; (d) Deltoid muscle. (Adapted from [29].)

the artifacts, it is possible to minimize them using two classical approaches: spatial filtering and some methods based on high-order statistical separation.

Spatial filtering is applied to raw EEG signals to reduce the correlation between channels caused by the spreading of the signals from the cortical source to the scalp and, therefore, improve the reliability of the BCI. Four spatial filters, namely a standard ear reference, a CAR, a small Laplacian, and a large Laplacian, will be discussed here. Figure 1.25 shows these four spatial filters in relation to the electrode C3 (highlighted in solid gray). For CAR and Laplacian filters, the mean activity of electrodes in black is subtracted from the activity recorded in C3.

As mentioned in Section 1.8, the CAR filter functions are based on the assumption that external biological artifacts affect all EEG electrodes in approximately the same way, due to the considerable distance from the noise source to the electrodes on the scalp. Therefore, the CAR method requires the subtraction, sample to sample, of the signals from the channels obtained from a common reference point, e.g., the ear. As the common average prioritizes signals that are present in a large number of channels, the subtraction leads to the elimination of these signals, working as a spatial filter [18].

The small Laplacian method also works as a spatial filter, requiring the subtraction, sample to sample, of the signals from the channels in a neighborhood with a radius of 3 cm around the analyzed channel, also obtained from a common reference point, e.g., the ear. The large Laplacian method works in the same way as the small Laplacian, requiring the subtraction, sample to sample, of the signals from the channels in a neighborhood with a radius of 6 cm around the analyzed channel, and also works as a spatial filter.

The study of spatial filters presented in Ref. [30–31] concluded that the CAR and large Laplacian methods are suitable for BCIs based on the paradigm of motor imagery. The motor cortex has radius between 6 and 12 cm, and then, Laplacian filters on the motor cortex with radius smaller than 6 cm would attenuate the information from the motor cortex itself.

As the spatial filters cannot eliminate all artifacts components, some methods for avoiding, detecting, and discarding or minimizing the EOG and eye blink artifacts will be discussed here. The first consideration for avoiding EOG artifact is to provide a fixation point during the mental tasks that helps to avoid eye movement artifacts. For example, an efficient strategy is to instruct subjects, during their mental tasks, to observe a cross in the center of the screen [13], in order to prevent eye movements.

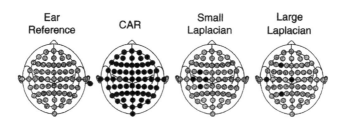

FIGURE 1.25   Electrode location for the spatial filters [30].

After taking care to minimize the artifacts during the experiments, the simplest approach to deal with the artifacts is to perform a threshold analysis. For example, whenever EEG signals from selected frontal channels exceed ±50 μV, a 0.5 s time window centered around the EEG peak of these electrodes is discarded in all EEG channels [24,25]. Another simple alternative using the threshold analysis is to replace the identified artifact window by a filtered EEG. For example, it is possible to replace the identified artifact window by a filtered window using a finite impulse response (FIR) equiripple high-pass filter set to 5 Hz, with forward and reverse order filtering algorithm to cancel the effect of phase distortion [14].

Figure 1.26 shows the same trial shown in Figure 1.23, which is contaminated by five eye blink artifacts. It shows the raw EEG, the application of the CAR filter, and the discard of artifacts by using the threshold analysis and the filtering method.

It can be seen that using the threshold approach, the eye blink artifacts are completely discarded in channel Fp1. It can be also noted in channel C3 that using the threshold and the filtering approaches, a peak of activity around 10 Hz becomes evident between 15 and 25 s, and this peak of activity lasts longer for the filtering approach, as the threshold approach causes loss of information. As mentioned earlier, this peak of activity in around 10 Hz represents an ERS occurring at μ band, which is a feature of interest for the identification of the mental task.

Some more complex approaches include performing blind source separation (BSS) using high-order statistics (HOS) methods, such as independent component analysis (ICA) [13,24,25,32]. In summary, the BSS problem is to perform the separation of a set of signals from a set of mixed signals, without the aid of information, or with very little information, about the source signals or the mixing process. BSS relies on the assumption that the source signals are not correlated with each other. Here, the mixed signals correspond to the EEG signal mixed with an unknown noise, and the BSS problem is related to perform the separation of a set of source signals, i.e., EEG artifact-free signals, from the set of mixed signals. ICA is a method for solving the BSS problem, recovering N independent source signals (s) from N linear mixtures (x), which relies on the assumption that the source signals are stationary and mutually statistically independent or decorrelated, whereas their mixtures are not. Statistical independence requires that all second-order and high-order correlations are zero, whereas decorrelation only seeks to minimize second-order statistics,

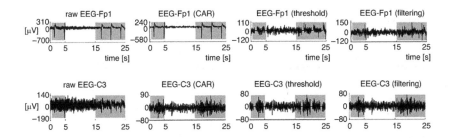

FIGURE 1.26 EEG of channels Fp1 and C3 during the application of the following cases: raw EEG; CAR method; threshold analysis and filtering.

which are the covariance or correlation. The linear mixture of N sources, with M samples each one, is composed of EEG signals recorded at different channels and at different time points. The columns of the source matrix (s) contain the time course of ICA components, and the columns of A give the relative projection strengths of the respective ICA components at each EEG site.

The scalp weights represent the fixed scalp topography of each ICA component, which provides evidence for the component physiological origin. For example, EOG and blink artifacts should project mainly to frontal sites, and then, the ICA components with high weight projected to frontal sites should be related to this kind of artifact, which can also be verified in the time course of the component [25,32,33].

Then, the ICA components related to artifacts can be set to zero, and artifact-free signals can be obtained projecting non-artifactual ICA components back onto the scalp. The artifact-free EEG signals can be obtained from the artifact-free source matrix and the mixing matrix A [25,32]. Figure 1.27 shows the ICA components and their fixed scalp topography of the same trial shown in Figures 1.23 and 1.26. This experiment corresponds to the mental task of imagination of movement of the right hand, which occurred between 5 and 15 s, with 19 electrodes positioned according to the International 10–20 System (Figure 1.17). Then, the input matrix has 19 linear mixtures, and ICA algorithm results in 19 independent components and, respectively, 19 scalp topographies. In the scalp topography, the relative strength of the ICA component on the 19 scalp sites is shown in shades of gray, in which light shades indicate high strength values (positive or negative), and dark shades indicate near-zero strength values.

Figure 1.27 shows that the first ICA component (ICA-1) is very similar to the time course of the eye blink artifact shown in the EEG of channel Fp1 in Figures 1.23 and 1.26. The scalp topography of this component is shown at its right side, and it can be seen that its strength is higher for frontal sites, in particular at Fp1. Then,

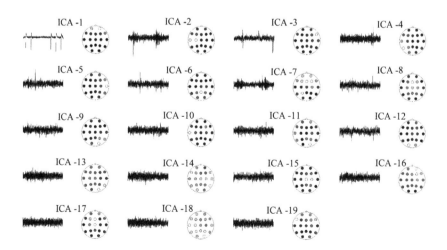

FIGURE 1.27 ICA components obtained by "fastICA" algorithm and their fixed scalp topographies.

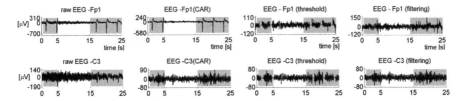

FIGURE 1.28 EEG of channels Fp1 and C3 and their periodograms during the application of the following cases: raw EEG; "fastICA" algorithm; and "runICA" algorithm.

this component can be set to zero to obtain a relative artifact-free EEG. Figure 1.28 shows the minimization of eye blink artifacts of the trial of Figure 1.23 by using two ICA algorithms, which are the "fastICA" and the "runICA". It can be seen that for both algorithms, the eye blink artifacts are completely discarded in channel Fp1, and there are no low-frequency artifacts. In channel C3, a peak of activity around 10 Hz can be noted, which becomes evident between 15 and 25 s, although "runICA" clearly distorts the original EEG amplitude.

The two mentioned ICA algorithms are based on the two broadest definitions of independence for ICA that are the maximization of the non-Gaussianity and the minimization of the mutual information (MMI) [13,24,25,32]. The non-Gaussianity family of ICA algorithms, which includes "fastICA" algorithm, is based on measures of non-Gaussianity such as kurtosis[30] and negentropy [34]. As the Gaussian distribution has above second-order cumulants equal to zero, the functions of its third- and fourth-order cumulants, such as the skewness[31] and kurtosis, are also zero, and then, they are used as a measurement for non-Gaussianity. The negentropy, or differential entropy, can also be approximated by a function of the first- and fourth-order cumulants, and then, it is also used as a measurement of non-Gaussianity. The "fastICA" is a fixed point algorithm, which, maximizing the absolute value of the kurtosis, leads to the identification of non-Gaussian sources [33].

The MMI family of ICA algorithms, which includes "runICA" algorithm, uses measures such as Kullback–Leibler divergence (KLD)[32] and maximum entropy. The "runica" performs ICA decomposition of input data using the logistic "infomax" algorithm described by Bell and Sejnowski in 1995 [35]. The "infomax" is an optimization principle in which a set of input values (I) are mapped to a set of output values (O) that are chosen, or learned, to maximize the average Shannon mutual information between the input and the output, $H(O;I)$. The mutual information can also be understood as the expectation of KLD of the conditional distribution of O given I, $p(o|i)$, and the univariate distribution of O, $p(o)$. Then, $H(O;I) = E[K_{LD}(p(o|i) \| p(o))]$, and the more different the distributions $p(o|i)$ and $p(o)$, the greater the information gain.

---

[30] The kurtosis, from the Greek word "kurtos", meaning curved, is a measure of flatness of a probability distribution.

[31] The measure of skewness of a probability distribution is given by, $\gamma_1 = k_3/k_2^{3/2}$, where k_i is the ith cumulant of the probability distribution.

[32] The Kullback–Leibler divergence, or information divergence, is a non-symmetric measure of the difference between two probability distributions P and Q, denoted $K_{LD}(P \| Q)$.

1.11.1 EVENT-RELATED POTENTIAL (ERP)

The EEG signals are composed of basic components of spontaneous potentials, which may be present throughout the range of frequencies of the EEG signal and are not produced by sensory stimulation. ERP is the change of the EEG potential in response to a particular event. ERP has much lower amplitude than the spontaneous activity, so that it cannot be recognized in the raw EEG. Therefore, average techniques are commonly employed for detecting the ERP. In the average technique, the ERP is considered to occur with an approximately constant delay in relation to the event, and the spontaneous activity is modeled as an additive random noise (Figure 1.29a) [37]. The EEG recordings obtained by repeating the same experiment or trial, under the same conditions, are called epochs, and as the number of epochs, N, used in the calculation of the average increases, the time-locked activity increases and the spontaneous activity decreases, and thus, the ERP can be observed.

The signal-to-noise ratio (SNR) is defined as the ratio of the average power in the signal to the average power in the noise [38]. As more epochs are used, the SNR of the time-locked event increases, allowing the observation of the ERP. In the ideal case, it is assumed that the measured EEG signals are made of a sequence of event-locked ERPs with invariable latency and shape, in addition to a noise, which can be approximated by a zero-mean Gaussian random process that is uncorrelated between trials and not time-locked to the event. The average power of the ERP signal is given by the expected value of its energy. When a signal is a stationary stochastic process, its power is defined to be the value of its correlation function at the origin. As the noise is supposed to be a stationary zero-mean Gaussian random process, its mean and variance do not vary with respect to time. Then, the correlation function at the origin is equal to its variance. The SNR of the EEG increases proportionally to the number of trials, and an excessive number of epochs will not result in significant changes in the ERP curve.

ERPs were originally called evoked potentials (EPs) because they are electrical potentials that are evoked by stimuli, as opposed to the spontaneous EEG rhythms. Concerning the terminology, Herb Vaughan wrote: "Since cerebral processes may

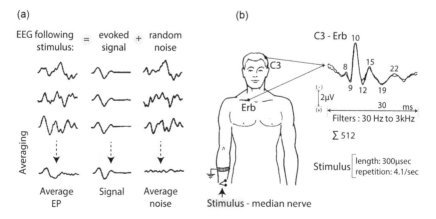

FIGURE 1.29 (a) ERP observation by averaging EEG recordings [37]; (b) normal somatosensory EP. (Adapted from [36].)

be related to voluntary movement and to relatively stimulus-independent psychological processes, the term evoked potentials is no longer sufficiently general to apply for all EEG phenomena related to sensory motor processes. Moreover, sufficiently prominent or distinctive psychological events may serve as time references for averaging, in addition to stimuli and motor responses. The term event-related potentials (ERP) is proposed to designate the general class of potentials that displays stable time relationships to a definable reference event" [10].

Then, the ERP may be related to voluntary movements or to relatively stimulus-independent psychological processes, such as the mental tasks. Depending on the modality to be studied, somatosensory, visual or auditory stimuli can be used, which have an important clinical utility for detecting neural degenerative diseases and traumatic pathologies [36]. EPs can be measured on specific areas of the cortex when certain nuclei of the thalamus are stimulated, which are called specific thalamic nuclei. Among them, e.g., there are the VPL nucleus and the MGN of the thalamus (Figure 1.11), whose stimulation evokes potentials, respectively, in the somatosensory and auditory areas of the cortex [5].

As mentioned above, EPs and ERPs are observed by means of hundreds of EEG recordings made with the same type of stimulus to eliminate the effects of random noise and enhance the event-locked response. Then, a temporal pattern related to the event becomes evident in the average EEG. The right part of Figure 1.29 shows a normal somatosensory EP, wherein the first high-voltage negative potential, with a latency of 10 ms, corresponds to the AP of the stimulated nerve (Erb). The numbers on the peaks of the EP curve indicate the latencies in milliseconds, and the number in the summation indicates that the potential of the figure is the result of the average of 512 EEG recordings [36].

1.12 MOVEMENT-RELATED (CORTICAL) POTENTIALS (MRPs/MRCPs)

MRCPs are ERPs measured on the motor cortex that have about 1 µV and are generated in response to an intention, or volition, to move a limb [37]. One kind of MRCP is the Bereitschafts potential (BP), also called readiness potential (RP) (Figure 1.30). The BP can be roughly identified in single-trial measurements, which is widely used in BCI applications. The BP is related to the premotor planning of volitional movement occurring in the motor cortex (M1) and SMA. The BP amplitude is ten to hundred times smaller than the α rhythm amplitude; then to observe its details, it is necessary to perform averaging. BP has two components: the early one (BP1), lasting from about -1.2 to -0.5 s; and the late component (BP2), lasting from -0.5 to shortly before 0 s.

The BP study has precipitated an interesting worldwide discussion about free will, which lasts for almost 40 years. Benjamin Libet studied in the 1980s the relationship between conscious experience of volition and the BP, finding that the BP starts about 0.35 s earlier than the subject has reported conscious awareness of the desire to make the movement. Libet's experiments suggest that the true initiator of volitional acts are some unconscious processes in the brain; therefore, free will plays no part in its initiation. Since the subjective experience of the conscious will to act precedes the action by only 200 ms, this leaves consciousness only 100–150 ms to veto an action

(Figure 1.30). This occurs because the final 50 ms prior to an act is occupied by the activation of the spinal motor neurons by the primary motor cortex. Libet concluded that we have no free will in the initiation of our movements, but, since we are able to prevent intended movement at the last moment, we do have the ability to veto these actions, sometimes called the "free won't" [39].

Soon et al. [40] performed an experiment with 36 right-handed subjects that should freely decide to move their left or right index finger at any time; meanwhile, their brain activity was measured using fMRI (functional magnetic ressonance imaging). Most of intentions (88.6%) were reported to be consciously formed 1 s before the movement, and support-vector machines (SVMs) were trained with fMRI data of several brain areas to predict the specific outcome of a subject's motor decision. The study showed that the activity of frontopolar and parietal cortex allowed the prediction of the finger laterality (left/right) 8 s before the action and allowed the time prediction (when the action would occur) 6 s before the action. They concluded that the earliest predictive information is encoded in frontopolar and parietal cortex, not in SMA, suggesting that the subjective experience of freedom is no more than an illusion and that our actions are initiated by unconscious mental processes long before we become aware of our intention to act [40]. Hans Helmut Kornhuber and Luder Deecke, who first recorded and reported the BP in 1964 [41], advocate that we have free will for the initiation of our movements and actions, but it is no absolute freedom, which would mean freedom from nature, considered impossible, as it is a relative freedom that was mentioned as freedom in "degrees of freedom".

In contrast to most ERPs and the BP, some brain events may not be observed in the EEG signal by using a simple linear technique such as the average over the time. Some events are restricted to some EEG rhythms, which are called phase-locked events. The relative decrement and increment of energy that occur in specific frequency bands are called event-related desynchronization (ERD) and event-related synchronization (ERS), respectively [6]. During the mental task, an ERD can be observed, which is followed by an ERS when the mental task is over. The events related to ERD/ERS and how they are measured is the topic of the next section.

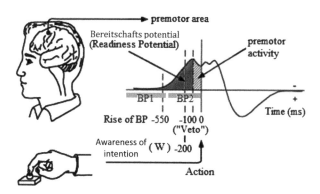

FIGURE 1.30   BP and the Libet's experiment. At −550 ms: the rise of BP; at −200 ms: the awareness of intention; at −100 ms: the possibility of veto of the action; at 0 s: the action of pressing a button.

1.13 ERD/ERS

As discussed earlier, the EEG measures the activity of neural masses working in synchrony. Usually, when a population of cortical neurons is inactive, these neurons receive a thalamic rhythmic afference that keeps them synchronized; thus, the sum of these rhythmic signals has large amplitude. When a neural mass is activated by a stimulus or an intention, cortical neurons receive distinct and specific signals, losing synchrony, and the sum of these signals has lower amplitude, which causes an ERD [1].

The bottom signal of Figure 1.22 shows the attenuation of the α rhythm, observed on V1, when the eyes are opened. The attenuation of the α rhythm amplitude when the eyes are opened corresponds to an ERD; on the other hand, the increase in the amplitude of the α rhythm when the eyes are closed corresponds to an ERS. As discussed in Section 1.5, the motor imagery inhibits the synchronization of M1 with the rhythmic activity of thalamus, causing an ERD similar to the ERD observed on V1. The μ rhythm was described by Jasper and Andrews [44] as the "precentral α rhythm", as it occurs under similar conditions as the α rhythm, but in the precentral area, on the motor cortex. Then, the electrodes on the precentral area (C3, Cz, and C4) can measure an ERD in the EEG signal during motor mental tasks. Even subjects with limb amputations present an ERD during the motor imagery of the phantom limb.

The ERD generated during the performance of the motor imagery task of the hand is contralateral, and therefore, the mental task related to the right hand causes an ERD at μ band of M1 in the left brain hemisphere, which can be measured by the electrode C3. On the other hand, the motor imagery of the left hand causes an ERD at the μ band of M1 in the right brain hemisphere, which can be measured by the electrode C4. While performing the motor imagery of movement of the hands, the M1 hand area, as well as the adjacent areas SMA and PMA, is activated (Figure 1.12). R. Beisteiner et al. [44] showed that the DC potential difference between the electrodes C3 and C4 is higher during the motor imagery of the right hand than during the motor imagery of the left hand (Figure 1.31).

Another example shows the difference in ERD/ERS obtained from the cortical motor areas, through C3-F_Z and C4-F_Z electrodes, during button movements with the index finger of the right hand (the subject performed self-paced, brisk button-press with his right index finger each 10 s during 10 min). During the finger movements, the ERD value decreased around 50% for both contralateral (C3-F_Z electrodes) and ipsilateral region (C4-F_Z electrodes), as shown in Figures 1.32c and 1.33c. On the

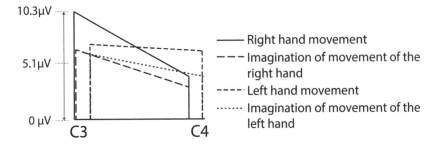

FIGURE 1.31 DC potential of electrodes C3 and C4. (Adapted from [44].)

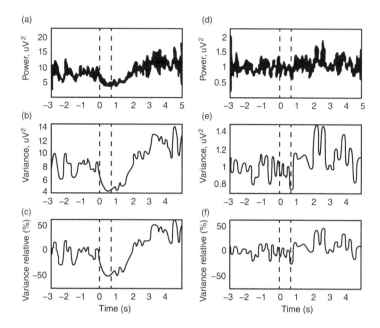

FIGURE 1.32 Representation of the ERD/ERS measured from the C3-F_Z electrodes. The vertical dotted lines represent the movement onset and offset. (a) Power spectrum smoothed of the ERD; (b) Variance inter-trial smoothed of the ERD; (c) Variance relative of the ERD; (d) Power spectrum smoothed of the ERS; (e) Variance inter-trial smoothed of the ERS; (f) Variance relative of the ERS.

other hand, the ERS value increased 50% after the end of the movement, only for the contralateral region (Figure 1.32c).

The left hemisphere of the brain is involved with the visual-spatial imagination, which is activated during the planning of mental tasks; thus, it has dominant contribution on the motor imagery. Then, the left hemisphere is somewhat active for the mental tasks of both hands and, by imagining the movement of the right hand, this potential is summed with the potential related to the activation of M1 that is also in the left hemisphere. While for the imagination of movement of the left hand, there will be a potential regarding the activation of the left hemisphere and the activation of M1 from the right hemisphere, then the potential difference between the electrodes C3 and C4 will be lower.

The classical method to measure the ERD/ERS is described by Gert Pfurtscheller and Fernando Henrique Lopes da Silva [6], and is similar to the process to obtain an ERP, but it includes filtering and squaring the signal. Given a set with multiple EEG recordings (epochs), the signals are filtered in the desired frequency range, and then the signals are squared obtaining the signal energy. Then, the average energy of all epochs is calculated to increase the SNR, making it possible to observe the ERD/ERS pattern. Here, the signal energy is used to prevent the cancelation of positive and negative EEG amplitudes during the average process. The curve obtained so far may be smoothed by using the moving average technique. Finally, as the ERD/ERS is

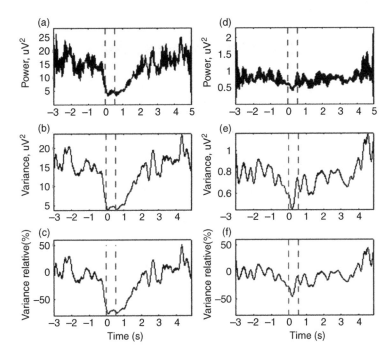

FIGURE 1.33 Representation of the ERD/ERS measured from the C4-F$_z$ electrodes. The vertical dotted lines represent the movement onset and offset. (a) Power spectrum smoothed of the ERD; (b) variance inter-trial smoothed of the ERD; (c) variance relative of the ERD; (d) power spectrum smoothed of the ERS; (e) variance inter-trial smoothed of the ERS; (f) variance relative of the ERS.

measured as percentages, the average energy of a previous reference period is calculated, and the relative signal energy can be measured in relation to a reference period.

Figure 1.34 shows the application of the classic method for measuring the ERD/ERS pattern. In this experiment, 60 epochs were used to observe the ERD/ERS, in which the subject performed the motor imagery task of the right hand. Figure 1.34a shows in gray the superimposition of the EEG from electrode C3 of the first, fifteenth, and thirtieth epochs, respectively, in black, dark gray, and light gray. The mental task lasted 10 s, occurring between 5 and 15 s of the EEG record, which corresponds to the central area not hatched of the figure. Figure 1.34b shows the EEG filtered at μ band. In this experiment, a FIR equiripple bandpass filter with forward and reverse order filtering algorithm is used to cancel the effect of phase distortion. This figure also shows the superimposition of the filtered EEG from the three aforementioned epochs. The objective of this figure is not to show individual characteristics of the epochs, but just to emphasize that the steps are performed individually, for each epoch.

In this experiment, the reference period is the one between 1 and 3 s of the recording, which is highlighted in Figure 1.34e as the interval shaded in dark gray. The horizontal line stresses the null value of the reference period. In Figure 1.34e, the energy decrease of around 70% occurring between 5 and 15 s is the ERD, and the energy

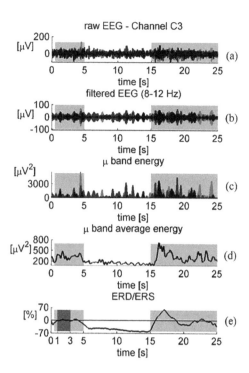

FIGURE 1.34 ERD/ERS calculation method. The reference period was taken between 1 and 3 s, and the mental task occurred between 5 and 15 s. (a) Raw EEG from channel C3 during the first, fifteenth, and thirtieth epochs, respectively, in black, dark gray, and light gray; (b) EEG from channel C3 filtered at μ band; (c) energy of μ band in channel C3; (d) average energy of μ band in channel C3 for 60 epochs; (e) ERD/ERS of μ band in channel C3.

increase of around 60% occurring between 15 and 20 s is the ERS. Therefore, the ERD/ERS pattern can be observed mainly at electrodes placed on the scalp region of the motor cortex during motor mental tasks.

As mentioned earlier, the ERD/ERS patterns of visual cortex can be directly observed on electrodes O1 and O2 through a frequency analysis without the need of averaging the signals. V1 is located on the occipital lobe, which is in the posterior region of the brain and roughly isolated from other lobes. Thus, the signals from V1 do not suffer much interference from signals of other brain areas, and consequently, the ERD/ERS pattern can be observed with a single EEG recording. However, the motor cortex is at the center of the brain surrounded by the frontal, temporal, and occipital lobes; thus, the signals generated by neighboring areas are mixed with signals from M1, with signals measured on electrodes C3, Cz, and C4 containing a mixture of signals from different brain areas. Therefore, the ERD/ERS pattern generated in M1 during the motor imagery of a limb cannot be observed with a single EEG recording, so it is necessary to calculate the average over a number of epochs, as described in Ref. [6].

It is worth mentioning that the procedure described above cannot be used for an online classifier, as it is necessary to observe the ERD/ERS in at least one epoch. A second method for obtaining the ERD/ERS is described in Ref. [45]. In fact, often, EPs

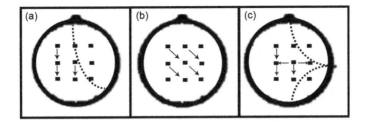

FIGURE 1.35 Activation and deactivation flow: (a) programming phase; (b) execution phase; (c) recovery phase. (Adapted from [46].)

spontaneously occurring in the brain in response to external stimuli or noise may mask any ERD/ERS, which occurs when the EP activity is in the same frequency band of the MRCP. The "inter-trial variance" (IV) method was then proposed to solve this problem.

Likewise, ERD and ERS are defined as the percentage values of increment or decrement in relation to a reference period. In this case, the reference period is formed by the sample variances of a period preceding the event. Bianchi et al. [46] analyzed the relations of coherence between EEG signals of nine electrodes (F3, Fz, F4, C3, Cz, C4, P3, Pz, and P4) filtered at α and β bands, and also analyzed the description of the time versus signal frequency of EEG through an autoregressive (AR) bivariate model, confirming the activation flow from the frontal area to the parietal area, through the direct pathway of movement, during the preparation of the movement (Figure 1.35). In this study, the ERD/ERS was calculated from the average of 80 epochs of index finger movements of the right hand, and the activation flow was obtained by analyzing the phase of the EEG power spectral density (PSD).

It can be noticed from Figure 1.35a, during the programming phase of the movement, that the activation flow involves only the frontal and parietal lobes of the left brain hemisphere, encompassing the electrode F3 to electrode P3. Thus, this shows the importance of using electrodes placed not only on the motor cortex, but also on the frontal and parietal lobes.

1.14 STEADY-STATE VISUAL EVOKED POTENTIALS (SSVEPs)

In addition to ERD/ERS, the other typical input patterns of BCIs are neurophysiological phenomena such as slow cortical potentials (SCP), P300, and steady-state visual evoked potentials (SSVEPs) [12,47].

An EP is the electrical response recorded from the human nervous system following presentation of a stimulus that can be detected by an EEG device. Transient visual evoked potentials (TVEPs) refer to electrical potentials initiated by brief visual stimuli, which are recorded from the scalp overlying the visual cortex. These responses occur when a subject observes a visual stimulus, such as a flash of light or a pattern on a monitor. TVEPs are used primarily to measure the functional integrity of the visual pathways from the retina via the optic nerves to the visual cortex of the brain. Their waveforms are usually extracted from the EEG signals by averaging. As shown in Figure 1.36, TVEP waveforms are represented using amplitude and time (latency) measurements [59].

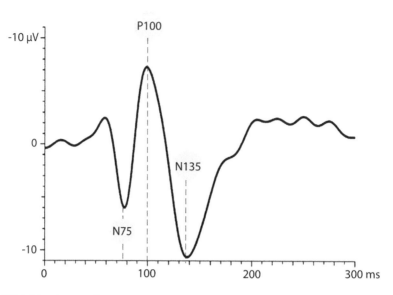

FIGURE 1.36 A normal pattern TVEP measured in EEG. (Adapted from [59].)

A TVEP waveform is obtained when the stimulus rate is low and the response is recorded over one single stimulus cycle. A typical TVEP waveform consists of N75, P100, and N135 peaks, which occur at about 75, 100, and 135 ms after visual stimulation, respectively. Any abnormality that affects the visual pathways or visual cortex in the brain can affect the TVEP waveform.

Unlike TVEP, SSVEP is the elicited response in the brain (mainly in visual cortex) by light stimuli flickering at a constant frequency. These potentials manifest as an oscillatory component in the EEG signal with the same frequency (and/or its harmonics) of the visual stimulation [56]. SSVEP can normally be evoked up to 90 Hz [57], and three stimuli bands can be identified: low (<12 Hz), medium (12–30 Hz), and high frequencies (>30 Hz) [56,58]. Whereas SSVEPs evoked by low- and medium-frequency stimuli bands are easier to detect (due to their higher energy), stimuli in these bands can produce epileptic seizures, which occur typically from 15 to 25 Hz [64], false positives due to α rhythm (8–13 Hz) [56,65], migraine headaches [66], and visual fatigue [66]. Thus, a suitable stimuli band is the high-frequency band; however, its evoked SSVEP is harder to detect, and literature has reported that approximately only 65% of people are able to operate a BCI based on high-frequency SSVEP [7], meaning that about 35% of people cannot achieve effective control of this kind of BCI system.

The steady-state potentials are distinguished from transient potentials by their constituent discrete frequency components, which remain relatively constant in amplitude and phase over a long time period [22,58]. Consequently, the amplitude distribution of the spectral content of SSVEP, with characteristic SSVEP peaks, remains stable over time (Figure 1.37). As these characteristics are constant, many applications can be derived from SSVEP properties.

SSVEP has some advantages for the development of a BCI, such as the following: low BCI illiteracy [7,48,49,51], requires few or no training [22,50,52,53,54],

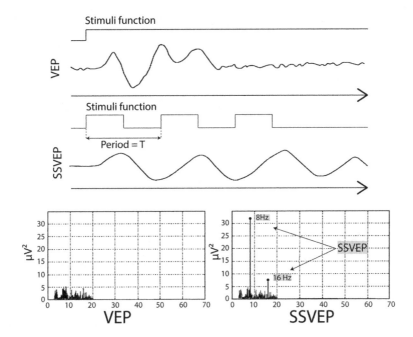

FIGURE 1.37 Example of a transient VEP and SSVEP response in time domain and frequency domain. (Adapted from [22].)

and presents high SNR, and high information transfer rate (ITR) [52], which is a standard measure of the amount of information transferred per unit of time (it will be detailed in Chapter 3). In SSVEP-based BCIs, the system options can be codified into visual stimuli (Figure 1.38), in which each stimulus oscillates at a specified frequency [55].

Thus, when a subject gazes at one of the stimuli, an SSVEP is evoked in their brain [22,60,61], which can be detected in the EEG signal. These measurements can then be used as control commands to the BCI with precision [48]. Figure 1.39 shows a diagram of an SSVEP-based BCI.

Three types of stimuli have mainly been used so far for SSVEP-based BCIs (Figure 1.40) [56]: (a) light stimuli (blinking light-emitting diodes – LEDs); (b) simple graphics stimuli, such as flickering squares on an LCD computer screen; (c) complex graphics flickers (e.g., alternatively reversing checkerboards).

Regarding a comparison of SSVEP efficiency using stimuli by LEDs and by a checkerboard in an LCD, studies conducted in Ref. [12] concluded that stimuli by LEDs, in low frequencies, allow detecting SSVEP, which is easier than stimuli by LCD. However, the ITR is higher when using stimuli by LCD. On the other hand, the same authors conducted studies about colors of stimuli to be used in SSVEP, analyzing red, green, blue, and yellow stimuli colors and concluding that the green color (followed closely by blue) is the most suitable option for visual stimuli in SSVEP, as it was considered by volunteers as the most comfortable in addition to being safe (compared with red, which can evoke epileptic responses).

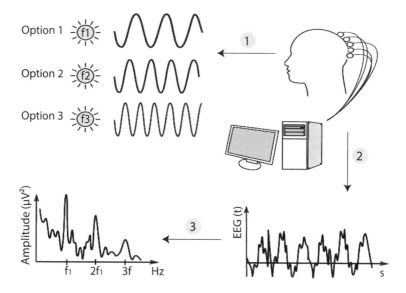

FIGURE 1.38　Design and operation of a SSVEP-based BCI. (1) Subjects are asked to attend a flickering stimulus. (2) Brain signals recorded during the stimulation. (3) The EEG signals are processed in order to extract representative features that are translated into commands. (Adapted from [55].)

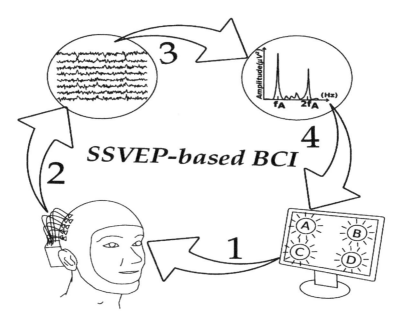

FIGURE 1.39　Diagram of the SSVEP-based BCI. (1) Subject gazes at target A flickering at frequency f_A. (2) EEG signals are measured from the scalp and recorded on a computer. (3) The data is processed, and features such as the peak at f_A and its harmonics are extracted. (4) The features are classified and translated into commands to the application.

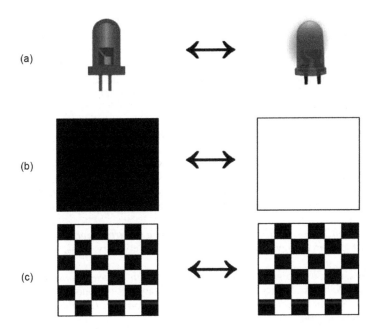

FIGURE 1.40 Main types of stimuli for SSVEP-based BCIs: (a) LEDs; (b) simple graphics stimuli; (c) complex graphics stimuli.

1.14.1 DEPENDENT AND INDEPENDENT SSVEP

Most of the existing BCIs based on SSVEP are considered "dependent", as these BCIs require the subject to perform eye gazing movements to focus on each SSVEP stimuli, thus generating a command. In contrast, SSVEP-based BCIs that are controlled by subject's focus of attention (FoA) are defined as "independent"; that is, the subject does not need to perform eye gazing movements, but only adjust their FoA to the stimulus of interest [62]. Independent SSVEP-based BCIs are important because possible end-users of BCI systems are patients at the end stage of amyotrophic lateral sclerosis (ALS) and Duchenne muscular dystrophy (DMD) in a situation called total locked-in syndrome (LIS), who may not control their eye movements and therefore might not be able to use dependent BCI systems.

Regarding colors of stimuli to be used in independent SSVEP, studies conducted in Ref. [63] with pairs of colored stimuli (white-black, green-red, and green-blue) found that the pair green-red elicits the highest SSVEP in the medium-frequency range (15–25 Hz), and green-blue stimulus elicits the highest SSVEP at high frequencies (30–40 Hz).

REFERENCES

1. Bear, M F; Connors, B W; Paradiso, M A, *Neurosciences: Unveiling the Nervous System*. In Portuguese, 3rd Edition, Artmed S.A., Sao Paulo, 2008.
2. Guyton, A C; Hall, J E, *Textbook of Medical Physiology*. 11th Edition, Elsevier Saunders, Philadelphia, PA, 2006.

3. Roland, P E; Larsen, B; Lassen, N A; Skinhøj, E, Supplementary motor area and other cortical areas in organization of voluntary movements in man. *Journal of Neurophysiology*, (43)1:118–136, 1980.
4. Hashemi, K, *The Source of EEG.* Available in: http://www.opensourceinstruments. com/Electronics/A3019/EEG.html. Accessed on: June 06, 2019.
5. Machado, A B M, *Neuroanatomia Funcional.* In Portuguese, 2nd Edition, Atheneu, Belo Horizonte, 2007.
6. Pfurtscheller, G; Lopes da Silva, F H, Event-related EEG/MEG synchronization and desynchronization: Basic principles. *Clinical Neurophysiology*, 110:1842–1857, 1999.
7. Huggins, J E;, Levine, S P J.; Fessler A; Sowers W M; Pfurtscheller G; Graimann, B; Schloegl A; Minecan D N; Kushwaha R K; Bement S L; Sagher, O; Schuh, L A, *Electrocorticogram as the Basis for a Direct Brain Interface: Opportunities for Improved Detection Accuracy.* First International IEEE EMBS Conference on Neural Engineering, 2003.
8. Davidson, R J; Jackson, D C; Larson, C L, *Human Electroencephalography. Handbook of Psychophysiology*, 2nd Edition, Cambridge University Press, New York, 27–52, 2000.
9. Lopes, C D, *EEG Signal Analysis Using the Discrete Wavelet Transform and Artificial Neural Networks.* Master Thesis, Federal University of Rio Grande do Sul (Brazil), 2005.
10. Luck, S J, *An Introduction to the Event-Related Potential Technique.* The MIT Press, Cambridge, 2005.
11. Tang, C; Fusheng, Y; Guang, C; Dakuan, G; Feng, F; Guosheng, Y; Xiuzhen, D, Correlation between structure and resistivity variations of the live human skull. *IEEE Transactions on Biomedical Engineering*, 55(9):2286–2292, 2008.
12. Silva, F L S; Niedermeyer, E, *Electroencephalography: Basic Principles, Clinical Applications, and Related Fields: Basic Principles, Clinical Applications and Related Fields.* 5th Edition, Lippincott Williams & Wilkins, Philadelphia, PA, 2012.
13. Ford, M R; Sands, S; Lew, H L, Overview of artifact reduction and removal in evoked potential and event-related potential recordings. *Physical Medicine and Rehabilitation Clinics of North America*, 15:1–17, 2004.
14. Hochberg, L R; Bacher, D; Jarosiewicz, B; Masse, N Y; Simeral, J D; Vogel, J; Haddadin, S; Liu, J; Cash, S S; Smagt, P; Donoghue, J P, Reach and grasp by people with tetraplegia using a neutrally controlled robotic arm. *Nature*, 485:372–375, 2012.
15. Aflalo, T; Kellis, S; Klaes, C; Lee, B; Shi, Y; Pejsa, K; Shanfield, K; Hayes-Jackson, S; Aisen, M; Heck, C; Liu, C; Andersen, R A, Decoding motor imagery from the posterior parietal cortex of a tetraplegic human. *Science*, 348(6237): 906–910, 2015.
16. Mulliken G H; Musallam S; Andersen R A, Decoding trajectories from posterior parietal cortex ensembles. *Journal of Neuroscience*, 28:12913–12926, 2008.
17. Bocker, K B E; Van Avermaete, J A G; Van den Berg-lenssen, M M C, The International 10–20 system revised: Cartesian and spherical coordinates. *Brain Topography*, 6:231–235, 1994.
18. Estebanez J M, *EEG-based analysis for the design of adaptive brain interfaces.* PhD Thesis, Centre de Recerca en Enginyeria Biomedica, Spain, 2003.
19. Thalamocortical Oscillations. *Scholarpedia.* Available in: http://www.scholarpedia. org/article/Thalamocortical oscillations. Accessed on: June 11, 2019.
20. Gonzalez, S L; Grave, R; Thut, G; Millan, J R; Morier, P; Landis, T, Very high frequency oscillations (VHFO) as a predictor of movement intentions. *NeuroImage*, 32(1):170–179, 2006.
21. Schaefer, R S; Vlek, R J; Desain, P, Music perception and imagery in EEG: Alpha band effects of task and stimulus. *International Journal of Psychophysiology*, 82:254–259, 2011.

22. St. Louis, E K; Frey, L C, *Electroencephalography (EEG): An Introductory Text and Atlas of Normal and Abnormal Findings in Adults, Children, and Infants.* American Epilepsy Society, Chicago, 2016.

23. Binnie, C; Cooper, R; Mauguire, F; Osselton, J; Prior, P; Tedman, B, *Clinical Neurophysiology*, Academic Press, Cambridge, 2003.

24. Croft, R J; Barry, R J, Removal of ocular artifact from the EEG: A review. *Clinical Neurophysiology*, 30:5–19, 2000.

25. Jung, T P; Makeig, S; Westerfield, M; Townsend, J; Courchesne, E; Sejnowski, T J, Removal of eye activity artifacts from visual event-related potentials in normal and clinical subjects. *Clinical Neurophysiology*, 111:1745–1758, 2000.

26. Grave de Peralta, M; Andino, S G; Perez, L; Ferrez, P W; Millan, J, Non-invasive estimation of local field potentials for neuroprosthesis control. *Cognitive Processing*, 6:59–64, 2005.

27. Cincotti, F; Mattia, D; Aloise, F; Bufalari, S; Astolfi, L; Fallani, F; Tocci, A; Bianchi, L; Marciani, M; Gao, S; Millan, J; Babiloni, F, High-resolution EEG techniques for brain-computer interface applications. *Journal of Neuroscience Methods*, 167:31–42, 2008.

28. Graimann, B; Huggins, J E; Levine, S P; Pfurtscheller, G, Visualization of significant ERD/ERS patterns in multichannel EEG and ECoG data. *Clinical Neurophysiology*, 113:43–47, 2002.

29. Grosse, P, *Diagnostic and Experimental Applications of Cortico-Muscular and Intermuscular Frequency Analysis.* Charite-Universitatsmedizin Berlin, Berlin, 2004.

30. Mcfarland, D J; McCane, L M; David, S V; Wolpaw, J R, Spatial filter selection for EEG-based communication. *Electroencephalography and Clinical Neurophysiology*, 103(3):386–394, 1997.

31. Benevides, A B; Bastos-Filho T; Sarcinelli-Filho M, Pseudoonline classification of three mental tasks using KL Divergence. *Journal of Medical and Biological Engineering*, 6:411–416, 2012.

32. Jung, T P; Makeig, S; Westerfield, M; Townsend, J; Courchesne, E; Sejnowski, T J, Analysis and visualization of singletrial event-related potentials. *Human Brain Mapping*, 14:166–185, 2001.

33. Vigario, R; Sarela J; Jousmaki, V; Hamalainen, M; Oja, E, Independent component approach to the analysis of EEG and MEG recordings. *IEEE Transactions on Biomedical Engineering*, 47(5):589–593, 2000.

34. Hyvarinen, A, Fast and robust fixed-point algorithms for independent component analysis. *IEEE Transactions on Neural Networks*, 10(3):626–634, 1999.

35. Bell, A J; Sejnowski, T J, An information-maximization approach to blind separation and blind deconvolution. *Neural Computation*, 7:1129–1159, 1995.

36. Delamonica, E A, *Electroencephalography.* In Spanish. El Ateneo, Buenos Aires, 1984.

37. Rugg, M D; Coles, M G H, *Electrophysiology of Mind Event-Related Brain Potentials and Cognition.* Oxford Psychology Series, Oxford, 1996.

38. Garcia, A L, *Probability and Random Process for Electrical Engineering.* 2nd Edition, Addison-Wesley Publishing Company, Reading, 1994.

39. Libet, B, Unconscious cerebral initiative and the role of conscious will in voluntary action. *Behavioral and Brain Sciences*, 8:529–566, 1985.

40. Soon, C S; Brass, M; Heinze, H J; Haynes, J D, Unconscious determinants of free decisions in the human brain. *Nature Neuroscience*, 11(5):543–545, 2008.

41. Deecke, L, *Experiments into readiness for action: 50th anniversary of the Bereitschaftspotential.* Available in: https://worldneurologyonline.com/article/experimentsreadiness-action-50th-anniversary-bereitschaftspotential/. Accessed on: June 18, 2019.

42. Nykopp, T, *Statistical Modelling Issues for the Adaptive Brain Interface*. Department of Electrical and Communications Engineering, Helsinki University of Technology, Espoo, 2001.
43. Smith, R C, *Electroencephalograph Based Brain Computer Interfaces*. Department of Electrical and Electronic Engineering, University College Dublin, Belfield, 2004.
44. Beisteiner, R; Hollinger, P; Lindinger, G; Lang, W; Berthoz, A, Mental representations of movements. Brain potentials associated with imagination of hand movements. *Electroencephalography and Clinical Neurophysiology*, 96:183–193, 1995.
45. Kalcher, J; Pfurtscheller, G, Discrimination between phaselocked and non-pahselocked event-related EEG activity. *Clinical Neurophysiology*, 94:381–384, 1995.
46. Bianchi, A M; Leocani, L; Mainardi, L T; Comi, G; Cerutti, S, Time-frequency analysis of event-related brain potentials. *Proceedings of the 20th Annual International Conference of the IEEE Engineering in Medicine and Biology Society*, 20(3):1486–1489, 1998.
47. HE, B et al., Brain-computer interfaces. *Neural Engineering*, 87–151, 2013.
48. Allison, B et al., BCI demographics: How many (and what kinds of) people can use an SSVEP BCI? *IEEE Transactions on Neural Systems and Rehabilitation Engineering*, 18(2):107116, 2010.
49. Volosyak, I, SSVEP-based Bremen-BCI interface-boosting information transfer rates. *Journal of Neural Engineering*, 8(3):036020, 2011.
50. Cheng, M et al., Design and implementation of a braincomputer interface with high transfer rates. *IEEE Transactions on Biomedical Engineering*, 49:1181–1186 2002.
51. Guger, C et al., How many people could use an SSVEP BCI? *Frontiers in Neuroscience*, 6:169, 2012.
52. Bin, G et al., An online multi-channel SSVEP-based braincomputer interface using a canonical correlation analysis method. *Journal of Neural Engineering*, 6(4):046002, 2009.
53. Vialatte, F B et al. Steady-state visually evoked potentials: focus on essential paradigms and future perspectives. *Progress in Neurobiology*, 90(4):418–438, 2010.
54. Ramadan, R A; Vasilakos, A V, Brain computer interface: Control signals review. *Neurocomputing*, 223:26–44, 2016.
55. Chumerin, N et al., Steady-state visual evoked potential based computer gaming on a consumer-grade EEG device. *IEEE Transactions on Computational Intelligence and AI in Games*, 5:100–110, 2013.
56. Zhu, D et al., A survey of stimulation methods used in SSVEP-based BCIs. *Computational Intelligence and Neuroscience*, 2010:702357, 2010.
57. Herrmann, C, Human EEG responses to 1–100 Hz flicker: resonance phenomena in visual cortex and their potential correlation to cognitive phenomena. *Experimental Brain Research*, 137(3–4):346–353, 2001.
58. Regan, D, *Human Brain Electrophysiology: Evoked Potentials and Evoked Magnetic Fields in Science and Medicine*. Elsevier, New York, 1989.
59. Odom, J V et al., Visual evoked potentials standard. *Documenta Ophthalmologica*, 108:115–123, 2004.
60. Norcia, A M et al., The steady-state visual evoked potential in vision research: A review. *Journal of Vision, The Association for Research in Vision and Ophthalmology*, 15(6):4, 2015.
61. Tello, R M G, *A novel approach of independent brain computer interface based on SSVEP*. PhD Thesis, Universidade Federal do Espírito Santo, 2016.
62. Allison, B Z et al., Towards an independent brain-computer interface using steady state visual evoked potentials. *Clinical Neurophysiology*, 119(2): 399–408, 2008.
63. Floriano, A S P, *Brain-computer interface based on high-frequency steady-state visual evoked potentials from below-the-hairline areas*. PhD Thesis, Universidade Federal do Espírito Santo, 2019.

64. Fisher, R S et al., Photic-and pattern-induced seizures: A review for the epilepsy foundation of America working group. *Epilepsia*, 46(9):1426–1441, 2005.

65. Yijun, W et al., *Brain-computer interface based on the high-frequency steady-state visual evoked potential.* First International Conference on Neural Interface and Control, Wuhan, China, 37–39, 2005.

66. Lin, F C et al., *High-frequency polychromatic visual stimuli for new interactive display systems.* The International Society for Optics and Photonics – SPIE, 2015. Available at: https://www.spie.org/news/5851-high-frequency-polychromatic-visual-stimuli-for-new-interactive-display-systems. Accessed on: Jan June 11, 2019.

67. Pomer-Escher, A; Souza, M; Bastos Filho, T F, *Analysis of Stress Level Based on Asymmetry Patterns of Alpha Rhythms in EEG Signals.* 5th IEEE Biosignals and Biorobotics Conference (BRC 2014), 2014.

68. Cotrina-Atencio, A; Garcia, J; Benevides, A; Longo, B; Ferreira, A; Pomer-Escher, A; Souza, M; Bastos Filho, T F, *Computing Stress-Related Emotional State via Frontal Cortex Asymmetry to be Applied in Passive-ssBCI.* 5th IEEE Biosignals and Biorobotics Conference (BRC 2014), 2014.

69. Delisle-Rodriguez, D; Villa-Parra, A C; Bastos-Filho, T F; Lopez-Delis, A; Frizera-Neto, A; Krishnan, S; Rocon, E, Adaptive spatial filter based on similarity indices to preserve the neural information on EEG signals during on-line processing. *Sensors*, 17:2725, 2017.

2 Brain–Computer Interfaces (BCIs)

*Alessandro Botti Benevides, Mario Sarcinelli-Filho,
and Teodiano Freire Bastos-Filho*

CONTENT

Brain–computer interfaces (BCIs) are systems that use the voluntary modulation of the neural activity to transmit information that may be used for communication or control. Nowadays, the neural activity can be monitored and translated into tractable electrical signals in either of two ways: electrophysiological and hemodynamic. The former refers to the interneuronal exchange of information through electrochemical transmitters to generate ionic currents flowing across neuronal assemblies [11]. Electrophysiological activity can be noninvasively measured by electroencephalography (EEG) [9,10,12,13] and magnetoencephalography (MEG) [14] or invasively measured by electrocorticography (ECoG) [15] or intracortical neuron recording [16,17]. Hemodynamic responses are described as processes to release glucose and oxygen through the bloodstream to active neural regions, which then create a local gradient of deoxyhemoglobin and oxyhemoglobin [18]. The changes in the local ratio can be measured and quantified by means of neuroimaging methods, such as functional magnetic resonance imaging (fMRI) and near-infrared spectroscopy (NIRS). Because hemodynamic responses are triggered by electrophysiological activity, they are only indirectly related to neuronal activity. This chapter shows the results of our researches conducted with BCIs based on EEG.

BCI is the union of two main processes: signal acquisition and signal processing (Figure 2.1). Regarding the signal acquisition, it is important to define the standard that is used for the placement of the electrodes on the scalp and the sampling rate of the system. The signal processing stage comprises signal preprocessing, feature extraction, pattern classification, and translation of the mental task to commands for a BCI application. The signal preprocessing stage is intended to reduce the amount of noise that contaminates the EEG signal, which usually employs spatial filters or high-order statistical (HOS) separation methods. The feature extraction stage focuses on finding the main features that differentiate the mental tasks, which can

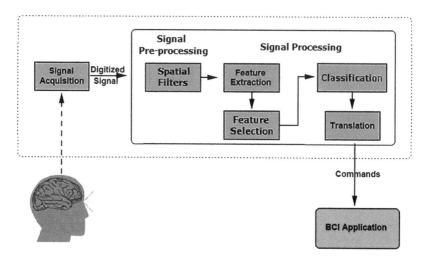

FIGURE 2.1   BCI timeline.

be followed by a feature selection step to refine the search for the best features. Then, the EEG features are sent to a classifier whose output is translated into commands to a BCI application.

BCIs were initially categorized by [26] as active BCIs, reactive BCIs, and passive BCIs. Active BCIs have outputs derived from brain activity, which are directly and consciously controlled by the user and thus independent of external events [28]. Reactive BCIs have outputs derived from brain activity, which arise due to reaction to external stimulation, which is indirectly modulated by the user [4]. Passive BCIs have outputs derived from implicit information on the actual user mental state, which arises arbitrarily without the purpose of voluntary control (actually, it fuses BCI technology with cognitive monitoring, providing the computer valuable information about the user's intentions, the situational interpretations, and mainly the emotional state). The first two categories derive their outputs for controlling an application, and the last one derives its output for improving human–environment interaction or human–machine interaction. Afterward, the concept of hybrid BCIs has been defined as a system composed of two BCIs, or at least one BCI and another system. According to [29], a hybrid BCI, like any BCI, must fulfill the following four criteria: (i) the device must rely on signals recorded directly from the brain, (ii) there must be at least one recordable brain signal that the user can intentionally modulate to effect the goal-directed behavior, (iii) real-time processing, and (iv) the user must obtain feedback.

Regarding the performance, a relevant issue in BCIs is the ability of efficiently converting user intentions into correct actions, and how to properly measure this efficiency. The traditional metrics to measure the efficiency of a BCI are classification accuracy, kappa, false-positive rate (FPR), true-positive rate (TPR), and information transfer rate (ITR) [22,23,24,25,28]. Accuracy (acc) is the percentage of correctly classified commands by the BCI. Kappa is a measure used for determining classification performance by removing the effect of accuracy of random classification, which can be calculated as [30]

$$\text{Kappa} = \frac{\text{acc} - 1/\text{NC}}{1 - 1/\text{NC}} \tag{2.1}$$

where NC is the number of classes.

FPR refers to the rate of incorrectly classified commands, whereas TPR refers to the rate of correctly classified commands, given by

$$\text{FPR} = \frac{\text{FP}}{\text{TN} + \text{FP}}, \quad \text{TPR} = \frac{\text{TP}}{\text{TP} + \text{FN}} \tag{2.2}$$

where FN and TN denote the number of false negatives and the number of true negatives, and FP and TP the number of false positives and the number of true positives, respectively.

ITR is a standard measure of the amount of information transferred per unit of time, which is defined as

$$\text{ITR} = s\left[\log_2(N) + p\log_2(p) + (1+p)\log_2\left(\frac{1-p}{N-1}\right) \right] \tag{2.3}$$

where N is the number of commands, p is the accuracy value, and s is the number of commands performed per minute. From (2.3), it can be seen that for a given number of N commands, the higher the accuracy, the higher the ITR.

BCI applications are employed in four major areas: (i) assistance for patients with severe motor disabilities, (ii) diagnosis of disorders of consciousness (DOC), (iii) entertainment applications, and (iv) recognition of affective or cognitive states.

i. **Assistance:** As mentioned in Section 1.14.1, BCI applications for assistance are mostly directed for patients in a situation called total locked-in syndrome (LIS),[1] or pseudo-coma. As the original motivation behind the development of BCIs is to help LIS patients, the next paragraphs briefly describe the types of LIS, its causes, the life quality, and life expectancy of these patients. In 1979, based on the extent of motor impairment, Bauer et al. [1] subdivided this syndrome into "classical LIS", characterized by total immobility, except for vertical eye movements or blinking; "incomplete LIS", in which there are remnants of voluntary motion; and "total LIS", which consists of complete immobility, including all eye movements, but with preserved consciousness.

Patients with incomplete and classical LIS may use generic human–machine interfaces (HMIs) adjusted in such a way that they can use devices as simple as a head pointer up to smart electronic interfaces controlled by their

---

[1] LIS was defined in 1966 by Fred Plum and Jerome Posner to describe the quadriplegia and anarthria resulting from the disruption of corticospinal and corticobulbar pathways, respectively, in brainstem damage.

biological signals, such as surface electromyography (sEMG) (e.g. muscle movements of the face or eye blinking) or electrooculography (EOG) (i.e., eyeball movements). However, patients with total LIS may solely use BCIs to communicate their thoughts to those around them by modulating their own neural activity. The Association of Locked-In Syndrome (ALIS) [2] shows that some sort of HMI is used by 81% of the LIS patients.

The most common causes of LIS are stroke or brain hemorrhage, traumatic brain injury, motor neuron diseases (such as ALS (amyotrophic lateral sclerosis) and DMD (Duchenne muscular dystrophy)), or medication overdose. The most common etiology of LIS is vascular pathology, either a basilar artery occlusion or a pontine hemorrhage. However, LIS may be caused by brainstem tumor, encephalitis, central pontine myelinolysis (CPM), vaccine reaction, and prolonged hypoglycemia [2]. Most LIS patients, with appropriate medical care, can return home, and their life expectancy is about several decades. Once the patient has medically stabilized in LIS for more than a year, 10-year survival rate is 83% and 20-year survival rate is 40% [2]. Even if the chances of motor recovery are very limited, computer-based communication methods, HMIs and BCIs have substantially improved the quality of life in chronic LIS [3]. Healthy individuals and medical professionals sometimes assume that the quality of life of a LIS patient is so poor that it is not worth living. On the contrary, chronic LIS patients typically self-report meaningful quality of life, and their demand for euthanasia is surprisingly infrequent. Among the causes of LIS death, just 10% are related to the patient's refusal of artificial nutrition and hydration [2].

Julia Tavalaro, Philippe Vigand, and Jean-Dominique Bauby are examples of LIS patients who maintained an active and productive life. In 1966, Tavalaro suffered two strokes and fell into a coma for 7 months. She was misdiagnosed with vegetative state (VS) for 6 years, until being considered with LIS in 1973. She wrote the book *Look Up for Yes* in 1997. In 1990, Vigand had a vertebral artery dissection and remained in a coma for 2 months. In 2000, he wrote *Only the Eyes Say Yes*, and he has written his second book *Dealing with the Menaced French Ecosystem* in 2002. Bauby had a brainstem stroke in 1995 and remained in coma for several weeks. He created the Association of Locked-In Syndrome (ALIS) and wrote the book *The diving bell and the butterfly*, which became a bestseller.

Regarding BCIs applied to severe LIS patients, it is worth noting the cases of Elias Musiris and Erik Ramsey. In 2002, Elias, a patient with chronic ALS, was the first total LIS person to regain some measures of communication through an EEG-based BCI, developed by the neurological researcher Niels Birbaumer. On the other hand, in 1999, 16-year-old Erik suffered a brainstem stroke after a car accident. He remained in coma for three weeks and was diagnosed with classical LIS. In 2004, he had a tiny electrode inserted into the part of the motor cortex responsible for the movements involved in speech. The electrode captured the ECoG signal of about 40 neurons, which were transmitted wirelessly under the scalp to a computer to be decoded, translated, and synthetized into single vowels.

ii. **Diagnosis:** A common pattern that can be observed in the aforementioned patients involves in detecting an early stage of coma. In fact, the misdiagnosis between VS and minimally conscious state (MCS) is around 40%, which leads to the second kind of application using BCIs, i.e., to improve the diagnosis of DOC [2,4,5,6]. This BCI application is receiving much attention because if VS is declared permanent, the ethical and legal issues about the withdrawal of treatment and organ harvesting can arise [2,5]. The next paragraphs describe the common progress of a victim of brain damage to exemplify how the VS and MCS arise. Some emphasis is given to the role of the EEG in the current diagnosis of brain death and the progress of using this type of signal to develop BCIs that can identify patients that are just minimally aware.

After a stroke or traumatic brain injury, the patient may go into a coma. Coma is a time-limited condition leading to either brain death, VS, or LIS, in rare cases. Brain death can be diagnosed with an extremely high rate of probability within hours to days of the original insult. The EEG test is the most validated and, because of its wide availability, preferred confirmatory test for brain death implemented in guidelines of many countries. The EEG test in patients with brain death shows an absence of ECoG activity, and the EEG signal becomes isoelectric, i.e., flat line, which is similar to a "functional decapitation". However, this EEG confirmatory test for brain death has sensitivity and specificity of around 90% [3].

The term "vegetative state" (VS) was defined in 1972 by Bryan Jennett and Fred Plum for patients with "wakefulness without awareness" of themselves or their environment. One month after the occurrence of brain damage, the VS is declared persistent, but does not mean that it is irreversible. The term "persistent vegetative state" (PVS) was defined in 1994 by the US Multi-Society Task Force on Persistent Vegetative State. VS may arbitrarily be regarded as PVS 3 months after a nontraumatic brain injury and 12 months after traumatic injury, and does imply that the patient will not recover. However, after being in VS, the patient can also progress to the MCS, which is an intermediate state between full awareness, as found in LIS patients, and no awareness at all, as found in the VS patients [2,3].

An important difference, though, is that patients who have remained in MCS for years still have a chance of recovery. In a much publicized case, Terry Wallis, who was considered to be in VS since a road accident in 1984, was actually in MCS for 19 years and started talking in 2003. Wallis also regained some ability to move his other limbs, although he cannot walk and still needs around-the-clock care [4,5]. Thus, the high misdiagnosis rate between VS and MCS may sound like a death sentence.

For example, in July 2005, a 23-year-old woman sustained a severe traumatic brain injury as a result of a road traffic accident. She remained comatose for more than a week and then evolved to VS for 5 months. Even if she fulfilled all of the criteria for a diagnosis of VS, according to international guidelines, the investigators conducted a second study in which they asked her to perform mental imagery tasks. When she was asked to imagine playing a game of tennis, the fMRI scans showed an activity in the supplementary motor area (SMA) of her brain, just as it did in healthy people. When she was asked to imagine walking through the rooms of her house, the scans showed the activation of the network involved in spatial navigation. Again, the response was indistinguishable from that seen in healthy people. Despite the clinical diagnosis that the patient was in VS, she understood the tasks and repeatedly performed them, and hence, it is believed that she was consciously aware of herself and her surroundings [4,5,7].

Similar studies are being conducted with EEG-based BCIs instead of fMRI, which has the advantage of being much cheaper, portable, and possible to use at the patient's bedside. For instance, Monti et al. [8] demonstrated that patients with DOC can use the modulation of their brain activity to reliably answer "yes" or "no" to simple questions, even though no signs of communication had ever been found through bedside examination. Bauer et al. [1] reported a case of a young comatose woman who failed to show any motor signs of conscious awareness at the intensive care unit. They performed a mental task to count a target name or her own name in a list of other names, and using EEG analysis, they were able to obtain the diagnosis of total LIS.

EEG studies measuring effective connectivity in α (8–12 Hz) and β (14–30 Hz) rhythms can differentiate between VS and MCS patients. Effective connectivity is a measure of the causal relationship between brain areas. Additionally, the EEG entropy was shown to be able to differentiate acute unconsciousness from MCS patients, with 89% sensitivity and 90% specificity. However, as the prognostic value of this measure is not high, it cannot be recommended as a prognostic tool [5].

Nevertheless, EEG is already very important to the confirmatory test of brain death and is also becoming important as a paradigm for single-switch BCIs (ssBCIs), which would improve the diagnostic reliability of VS and MCS. ssBCIs are simpler BCIs that operate with only two mental tasks, in which one of them is the absence of making any particular mental task, such as studied by [1,8]. These simpler BCIs are proving useful to confirm the patient awareness. In fact, a European collaborative project named EU-Decoder (2010–2013) [6] conducted studies to deploy ssBCIs for the detection of consciousness in nonresponsive patients and hence improve DOC diagnosis. This project used ssBCIs based on EEG and NIRS, both portable, with the aim of being applied at the patient's bedside.

The American Academy of Neurology (AAN) has published a position statement, in 1993, concerning that LIS patients have the right to make health care decisions about themselves, including whether to accept or refuse life-sustaining therapy, and either not start or stop once it started [5]. Thus, through BCIs, these patients can express their preferences in terms of treatment planning, like pain management or end-of-life decision-making [2].

iii. **Entertainment:** Entertainment applications are relatively a new branch in the area of BCIs, which are intended for healthy people who want to control games and devices through their brain waves [27]. Some companies that are engaged in BCI products such as interfaces, Software Development Kit (SDK), and games are Interactive Productline (www.mindballplay. com), OCZ Technology (www.ocztechnology.com), Emotiv (www.emotiv.com), NeuroSky (http://neurosky.com), EyeComTec (www.eyecomtec. com), InteraXon (choosemuse.com), and OpenBCI (openbci.com). For a short review, the Swedish company Interactive Productline developed the game Mindball, in which players compete to control a ball's movement across a table by modulating their brain waves, becoming more relaxed or focused. The manufacturer of computer hardware, OCZ Technology, developed a computer game controller, NIA (Neural Impulse Actuator), which is based on facial sEMG and EEG. The neuroengineering company Emotiv has brought to market a low-cost EEG interface called EPOC, which is able to play games specifically developed for it, use it to connect to a computer to play current games, or use it to be part of a BCI. NeuroSky is a BCI company that developed the EEG-based headset called Mindset and the compatible toys Mindflex and Star Wars Force Trainer, in which players lift a ball by concentrating, and afterward also launched

FIGURE 2.2 Child with ASD wearing a wireless EEG device to recognize his emotions while interacting with a social robot through an aBCI.

a headset called MindWave. EyeComTec sells the EEG-based headset XWave Sonic and the EEG-based headband XWave Sport for EEG use and analysis in computer systems. InteraXon commercializes the EEG headset Muse for meditation, and OpenBCI launched an open-source BCI (freely available software and hardware) and also sells caps and headsets for EEG acquisition.

iv. **Cognitive states:** Affective brain–computer interfaces (aBCIs) are an approach of affective computing directed to BCIs. Affective computing is a branch of computer science originated in 1995 related to the study and development of systems that can recognize, interpret, process, and simulate human emotions. Affective computing may use several kinds of features to perform the recognition of human affects, such as facial image, body gesture image, blood volume pulse, galvanic skin response, facial sEMG, and EEG. See an example of an aBCI developed in our laboratory at UFES/Brazil to recognize emotions of children with ASD (autism spectrum disorder) while interacting with a social robot (Figure 2.2) [19]. The common basic emotions that are used for recognition are anger, disgust, fear, happiness, sadness, and surprise.

Some aBCIs intend to perform the automatic recognition of attention or fatigue that would be used to adapt the BCI to the user state and thus improve the success rate of the BCI during the identification of the mental tasks, such as done by the hybrid BCIs developed in our laboratory applied to a robotic wheelchair and an autonomous car [20,21].

REFERENCES

1. Bauer G; Gerstenbrand F; Rumpl E, *Varieties of the locked-in syndrome. J. Neurol.*, 221(2):77–91, 1979.
2. Laureys, S; Pellas, F; Van Eeckhout, P; Ghorbel, S; Schnakers, C; Perrin, F; Berre, J; Faymonville, M E; Pantke, K. H; Damas, F; Lamy, M; Moonen, G, Goldman, S, The locked-in syndrome: what is it like to be conscious but paralyzed and voiceless?. *Prog. Brain Res.*, 150:495–511, 2005.
3. Laureys, S; Owen, A M, Schiff, N D, Brain function in coma, vegetative state, and related disorders. *Lancet Neurol.*, 3:537–546, 2004.
4. Laureys, S, Eyes open brain shut. *Sci. Am.*, 4:3237, 2007.
5. Gantner, I S; Bodart, O; Laureys, S; Demertzi, A, Our rapidly changing understanding of acute and chronic disorders of consciousness: challenges for neurologists. *Future Neurol.*, 8(1):43–54, 2013.
6. EU-DECODER: Deployment of Brain-Computer Interfaces for the Detection of Consciousness in Non-Responsive Patients. Available in: https://graz.pure.elsevier.com/en/projects/eu-decoderdeployment-of-brain-computer-interfaces-for-the-detect. Accessed on: June 5, 2019.
7. Owen, A M; Coleman, M R; Boly, M; Davis, M H; Laureys, S; Pickard, J D, Detecting awareness in the vegetative state. *Science*, 313:1402, 2006.
8. Monti, M M; Vanhaudenhuyse A; Coleman, M R; Boly M; Pickard J D; Tshibanda L; Owen A M; Laureys S, Willful modulation of brain activity in disorders of consciousness. *N. Engl. J. Med.*, 362:579–589, 2010.

9. Allison, B Z; Dunne, S; Leeb, R; MillA¡n, J; Nijholt, A, *Towards Practical Brain-Computer Interfaces: Bridging the Gap from Research to Real-World Applications.* Berlin: Springer, 2012.

10. Nicolas-Alonso, L F; Gomez-Gil J, Brain computer interfaces, a review. *Sensors,* 12(2):1211–1279, 2012.

11. Shepherd, G M, *Neurobiology.* New York: Oxford University Press, 1988.

12. Wolpaw, J R; Birbaumer, N; McFarland D, J; Pfurtscheller, G Vaughan, T M, Brain-computer interfaces for communication and control. *Clin. Neurophysiol.,* 113(6):767–791, 2002.

13. Rached, T S; Perkusich, A, Emotion recognition based on brain-computer interface systems. Brain-computer interface systems - Recent progress and future prospects, Intechopen, 2013.

14. Mellinger, J; Schalk, G; Braun, C; Preissl, H; Rosenstiel, W; Birbaumer, N; Kubler, A., An MEG-based brain-computer interface (BCI). *NeuroImage,* 36(3):581–593, 2007.

15. Schalk, G; Leuthardt, E C, Brain-computer interfaces using electrocorticographic signals. *IEEE Rev. Biomed. Eng.,* 4:140–154, 2011.

16. Santhanam, G; Ryu, S I; Yu, B M; Afshar, A; Shenoy, K V, A high-performance brain-computer interface. *Nature,* 442(7099):195–198, 2006.

17. Schwartz, A B; Cui, X T; Weber, D J; Moran, D W, Brain-controlled interfaces: movement restoration with neural prosthetics. *Neuron,* 52(1):205–220, 2006.

18. Jasdzewski, G; Strangman, G; Wagner, J; Kwong, K K; Poldrack, R A; Boas, D A, Differences in the hemodynamic response to event-related motor and visual paradigms as measured by near-infrared spectroscopy. *NeuroImage,* 20(1):479–488, 2003.

19. Goulart, C; Valadao, C; Caldeira, E; Freire-Bastos, T, Brain signal evaluation of children with Autism Spectrum Disorder in the interaction with a social robot. *Biotechnol. Res. Innovat.,* 1: 1–9, 2018.

20. Cotrina-Atencio, A; Benevides, A; Ferreira, A; Bastos-Filho, T; Garcia, J; Menezes, M; Pereira, C, Towards an Architecture of a Hybrid BCI Based on SSVEP-BCI and Passive-BCI. *36th Annual International Conference of the IEEE Engineering in Medicine and Biology Society (EMBC'14),* 2014.

21. Garcia, J F C; Muller, S M T; Caicedo, E; Bastos Filho, T F; Souza, A, Non-fatigating brain computer interface based on SSVEP and ERD to command an autonomous car. *Adv. Data Sci. Adap. Anal.,* 1:1–11, 2018.

22. Vialatte, F B et al., Steady-state visually evoked potentials: focus on essential paradigms and future perspectives. *Prog. Neurobiol.,* 90(4), 418–438, 2010.

23. Ang, K K; Chin, Z Y; Wang, C; Guan, C; Zhang, H, Filter bank common spatial pattern algorithm on BCI competition IV Datasets 2a and 2b. *Front. Neurosci.,* 6:39, 2012.

24. Delisle-Rodriguez, D; Villa-Parra, A C; Bastos-Filho, T F; Lopez-Delis, A; Frizera-Neto, A; Krishnan, S; Rocon, E, Adaptive spatial filter based on similarity indices to preserve the neural information on EEG signals during on-line processing. *Sensors,* 17:2725, 2017.

25. Barachant, A; Bonnet, S; Congedo, M; Jutten, C, Multiclass brain-computer interface classification by Riemannian geometry. *IEEE Trans. Biomed. Eng.,* 59:4, 2012.

26. Zander, T; Kothe, C, Towards passive brain-computer interfaces: applying brain-computer interface technology to human-machine systems in general. *J. Neural. Eng.,* 8(2): 025005, 2011.

27. Wolpaw, J; Neat, M D G; Forneris, C, An EEG-based brain-computer interface for cursor control. *Electroencephalog. Clin. Neurophysiol.,* 78(3):252–259, 1991.

28. Muller, S M T; Celeste, W C; Bastos-Filho, T F; SarcinelliFilho, M, Brain-computer interface based on visual evoked potentials to command autonomous robotic wheelchair. *J Med. Biol. Eng.*, 30(6):407–415, 2010.

29. Pfurtscheller, G; Allison, B Z; Brunner, C; Bauernfeind, G; Solis-Escalante, T; Scherer, R; Zander, T O; Mueller-Putz, G; Neuper, C; Birbaumer, N, The hybrid BCI. *Front Neurosci.*, 4:30, 2010.

30. Huang, W; Zhao J; Fu, W, A deep learning approach based on CSP for EEG analysis. Shi Z., Mercier-Laurent E., Li J. (eds) *Intelligent Information Processing IX. IIP 2018.* IFIP Advances in Information and Communication Technology, 538, Berlin: Springer, 2018.

3 Applications of BCIs

André Ferreira, Sandra Mara Torres Müller, Javier Ferney Castillo Garcia, Richard Junior Manuel Godinez-Tello, Alan Silva da Paz Floriano, Anibal Cotrina Atencio, Leandro Bueno, Denis Delisle Rodríguez, Ana Cecilia Villa-Parra, Alejandra Romero-Laiseca, Alexandre Luís Cardoso Bissoli, Berthil Longo, Alexandre Geraldo Pomer-Escher, Flávia Aparecida Loterio, Christiane Mara Goulart, Kevin Antonio Hernández-Ossa, Maria Dolores Pinheiro de Souza, Jéssica Paola Souza Lima, Vivianne Cardoso, Celso De La Cruz Casaño, Hamilton Rivera-Flor, Eduardo Henrique Couto Montenegro, Thomaz Rodrigues Botelho, Dharmendra Gurve, Jeevan Pant, Muhammad Asraful Hasan, and Sridhar Krishnan, Eliete Caldeira, Anselmo Frizera-Neto, Mario Sarcinelli-Filho, and Teodiano Freire Bastos-Filho

CONTENTS

This chapter presents the brain–computer interfaces (BCIs) developed in our laboratory at UFES/Brazil along 20 years of investigation.

3.1   BCI BASED ON ERD/ERS PATTERNS IN α RHYTHMS

3.1.1   COMMAND OF ROBOTIC WHEELCHAIR AND AUGMENTATIVE AND ALTERNATIVE COMMUNICATION ONBOARD SYSTEM

This BCI makes use of ERD (event-related desynchronization) and ERS (event-related synchronization) patterns in α rhythm (8 to 12 Hz) [1]. As discussed in Section 1.13, ERD is related to concentration or existence of visual stimulus (suppression), whereas ERS is related to relaxation with few stimuli or the absence of visual stimulus (activation) (Figure 3.1). In this application, the electroencephalogram (EEG) signals are captured on O1 and O2 locations. The technique used for feature extraction is the energy of α rhythm, and the classifier is based on the signal variance. This BCI reached a success rate of 100%.

This BCI was implemented onboard a robotic wheelchair (Figure 3.2), which has a smartphone located in front of the user's eye field, allowing the choice of icons to generate commands to either wheelchair movements, or augmentative and alternative communication (AAC).

For the wheelchair movements, the icons represent arrows or curves (discrete movement) or pictograms of home places (final destination), as shown in Figure 3.3. On the other hand, for AAC, the icons are letters (to compound sentences) or pictograms representing felling or needs (Figure 3.4), whose outputs are artificial voices emitted through speakers onboard the wheelchair. Before using this wheelchair, users can teleoperate it and the mental effort to use the BCI is evaluated through the NASA Task Load Index (NASA-TLX) [21]. The limitation of this BCI is that it depends on eye closing to generate the ERS patterns and thus choosing the wished icon.

With this BCI, it is also possible to turn home appliances on/off, change channels and increase/decrease the volume of TV set, and make mobile phone calls or send text messages (for pre-recorded numbers), from the wheelchair, through a wireless

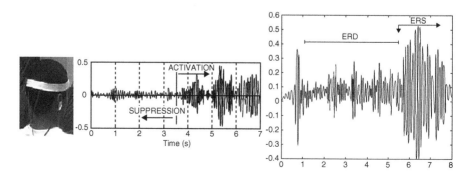

FIGURE 3.1   ERD and ERS patterns in α band.

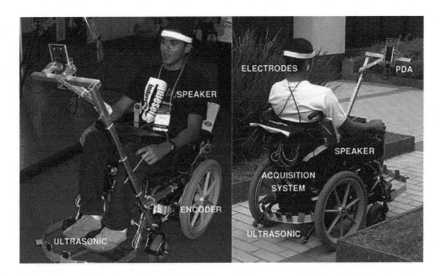

FIGURE 3.2 Robotic wheelchair with onboard AAC system commanded by ERD/ERS in α rhythms.

connection. In this case, the same system for smart home developed in Ref. [2] is used, which makes use of the Emotiv Epoch wireless EEG acquisition device (Figure 3.5), or rearranging that device to be used in other EEG caps positioned on the occipital region, and also with dry EEG electrodes [3] (Figure 3.6).

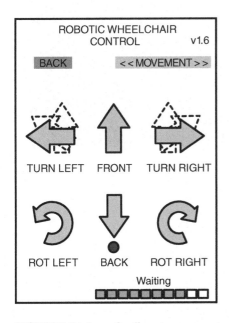

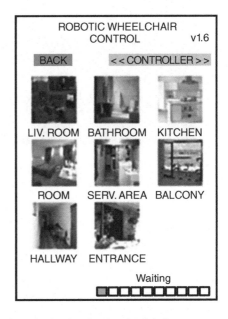

FIGURE 3.3 Icons for discrete movements or final destination for the wheelchair.

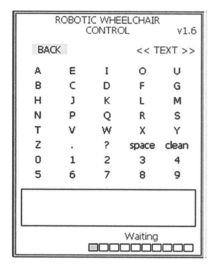

FIGURE 3.4   Icons to compound sentences or express feeling or needs.

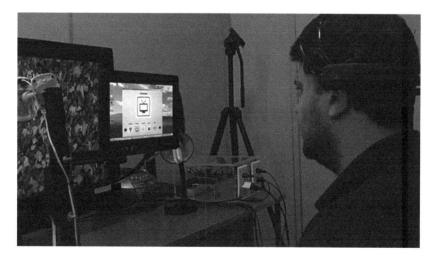

FIGURE 3.5   User using Emotiv Epoch to command, from his wheelchair, home appliances in a smart home.

3.2   BCIs BASED ON DEPENDENT SSVEP

3.2.1   Command of a Robotic Wheelchair

In this BCI, the stimuli screen is generated by field programmable gate arrays (FPGA), which consists of four stripes (checkboard) plus four LEDs used as visual feedback to motivate the user to command the wheelchair (Figure 3.7) [4]. The flickering frequencies used to command the wheelchair to go forward, left, backward,

FIGURE 3.6   EEG caps using the modified Emotiv Epoch device to be used on the occipital region and with dry electrodes.

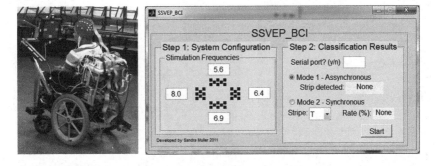

FIGURE 3.7   User commanding a wheelchair through focus on a stimuli screen with four flickering stripes. The LEDs on the screen borders are used as visual feedback to motivate the user regarding the correct detection of their commands.

and right are 5.6, 6.4, 6.9, and 8.0 rps (repetition per second), respectively. Twelve EEG channels were used, and the preprocessing included digital low-pass (fifth-order elliptic filter) and bandpass from 3 to 60 Hz filtering. A spatial filter based on common average reference (CAR) was also applied. The feature extraction was performed by spectral F-test, and the classifier was a rule-based classifier (decision tree). The success rate obtained with this BCI reached 96%, with ITR reaching 101.7 bits/min.

Another steady-state visual evoked potential (SSVEP)-based BCI [5] used high frequencies (37, 38, 39, and 40 rps) to command this robotic wheelchair, providing a more comfortable way of using SSVEP, as no (or very few) visual fatigue is generated for high frequencies, which is an advantage for long-term BCI operation. With this BCI, a success rate of 85% and ITR of 72.5 bits/min were reached. A limitation of both SSVEP-based BCIs is that they are dependent on eye gaze, which are classified as "dependent SSVEP-based BCIs".

3.2.2 BCI Based on SSVEP to Command a Telepresence Robot and Its Onboard AAC System

Other example of application is a BCI to command the movements of a telepresence robot and its onboard AAC system (Figure 3.8). This BCI allows a subject to remotely control a robot to displace in an environment and communicate with people around the robot (bidirectional communication of video/audio). In this case, the preprocessing is carried out by a digital low-pass (finite impulse response), CAR filter and bandpass from 3 to 20 Hz. The feature extraction is done by spectral F-test, and the classifier is the canonical correlation analysis (CCA). The success rate of this BCI was 88%, with an ITR of 13 bits/min. Such as the previous one, the limitation of this is that it is a "dependent SSVEP-based BCI".

FIGURE 3.8 User commanding, through SSVEP, a telepresence robot to move it around and establish a remote conversation using the onboard AAC system.

3.2.3 Hybrid-BCI Based on SSVEP to Command an Autonomous Car

Another application of the previous BCI plus the use of ERD/ERS patterns of α rhythms (hybrid-BCI) was the command of the autonomous car of UFES/Brazil [6]. The car model is an Escape Hybrid (Ford) equipped with accelerator control system, brake, gear, lights, horn, odometry, steering wheel and motor speed, laser system "LIDAR" (light detection and ranging), high-speed video cameras (firewire), GPS (Global Positioning System), and 6D IMU (inertial measurement unit) used for SLAM (simultaneous localization and mapping). In this case, the stimuli screens are into the car, which consist of two flickering stripes (Yes or No) (Figure 3.9).

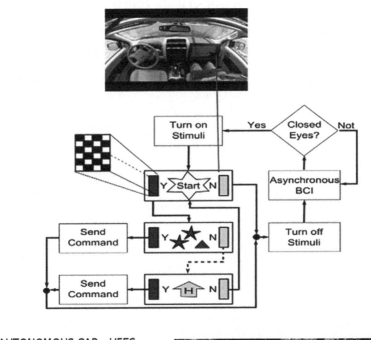

FIGURE 3.9 Command of the autonomous car of UFES/Brazil through SSVEP with two flickering stripes (Y or N) to be chosen according to the picture of the final destination sequentially exhibited on the screen.

The flickering frequencies associated with "Y-Yes" or "N-No" to choose the final destination to the autonomous car are 6.4 and 8.0 rps, respectively. This BCI reached a success rate of 96% and ITR of 13 bits/min. As low-frequency stimuli were used, an adopted strategy to decrease the visual fatigue and/or headache was that once the subject had chosen the final destination, he/she should close their eyes for a while (which affected the ERD/ERS patterns, as explained in Section 3.1.1), and this turned off the flickering stimuli. To turn them on again, another eye closing was necessary. This is a typical application of a hybrid-BCI, which improved the original SSVEP-BCI through the elimination of visual fatigue.

3.3 BCIs BASED ON INDEPENDENT SSVEP

This BCI [7,8] is based on depth-of-field (DoF) using SSVEP with stimuli by LEDs, which is used to command the telepresence robot movements (with bidirectional communication of video/audio) and its onboard AAC system (Figure 3.10). The flickering frequencies are 5.6 and 6.4 rps. Twelve EEG channels are used, and the preprocessing includes digital low-pass (finite impulse response), CAR filter and bandpass from 3 to 20 Hz filtering. For feature extraction, LASSO (least absolute shrinkage and selection operator) is used, and the classifier is a CCA. This BCI reached a success rate of 96% and ITR of 9.6 bits/min. An advantage of this BCI is that it does not depend on eye gaze, but only on change of focus.

This independent SSVEP-based BCI was also applied to command a robotic wheelchair using Rubin's face–vase stimuli (Figure 3.11) [9,10,12]. In this application, two LEDs (for vase and faces) are used as stimuli. The flickering frequencies are 15.0 Hz (vase) and 11.0 Hz (faces), and three EEG channels are used: O1, O2, and Oz. The feature extraction is performed by spectral F-test, and the classifiers are multivariate synchronization index (MSI) and CCA. The success rate for this BCI was 82.7% for "face" and 76% for "vase", and ITR was 35.18 bits/min.

FIGURE 3.10 Command of a telepresence robot, using an independent SSVEP based on DoF, to move it around and establish a remote conversation using the onboard AAC system.

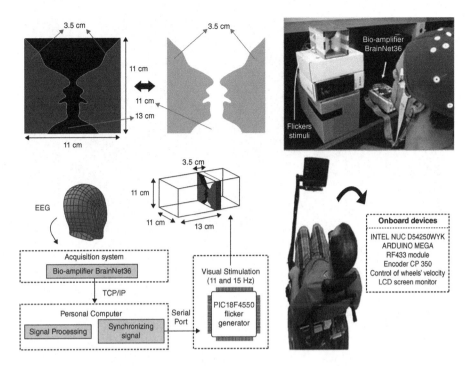

FIGURE 3.11 Command of a robotic wheelchair, using an independent SSVEP based on DoF, to move it around and establish a remote conversation using the onboard AAC system.

3.4 COMPRESSIVE TECHNIQUE APPLIED TO SSVEP-BASED BCI

When the EEG signal needs to be wirelessly transmitted, data compression may be necessary. A solution to compress the data is to use a technique named "compressive sensing", which is based on compressing EEG signals using random projection. This is an emerging and promising technique for the development of low-power, small-chip, and robust wireless BCIs, whose compressed data are transmitted through wireless connection. At the end, the EEG signals are reconstructed through l_p^d-regularized least squares, l_p^{2d}-regularized least squares, and block-sparse Bayesian learning bound optimization.

This solution was used in a SSVEP-based BCI [11], which uses minimum energy combination (MEC), CCA, and multivariate synchronization index (MSI) to detect the SSVEP, whose flickering frequencies are 8 Hz and 13 Hz (Figure 3.12). Five compression rates, i.e., 95%, 90%, 85%, 80%, and 75%, and a window length of 1 and 4 s were evaluated for this BCI.

An alternative technique for SSVEP detection was evaluated in the same BCI using correlation analysis between tensor models (PARAFAC – parallel factor analysis) [12] (Figure 3.13). In this technique, a three-way EEG tensor of channel versus frequency versus time is decomposed into constituting factor matrices using the PARAFAC model, which allows the decomposition of multichannel EEG into

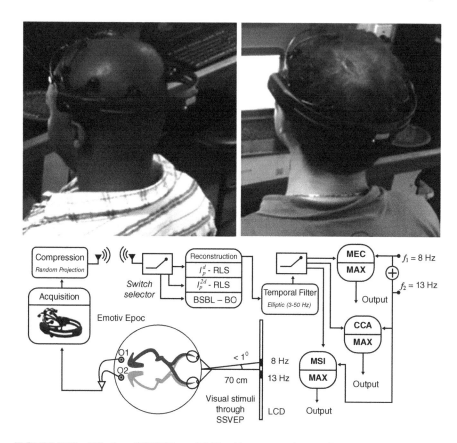

FIGURE 3.12   Wireless SSVEP-based BCI with compressive sensing.

constituting temporal, spectral, and spatial signatures. The success rate reached with this independent BCI was 83%, with an ITR of 21 bits/min.

3.5   BCIs BASED ON MOTOR IMAGERY

Several studies were conducted in our laboratory regarding the detection of the motor imagery, which can be done offline and online. Some of these offline BCIs used database from BCI competitions [13]. In Ref. [14], EEG signals captured from C3, C4, and Cz in such database were filtered from 8 to 13 Hz. Short-time Fourier transform (STFT) and support vector machine (SVM) were used for feature extraction and classifier, respectively. Three classes were considered: imagination of right-hand movements, left-hand movements, and noise. With this BCI, the success rate reached was 98%.

Another BCI was also proposed in Ref. [15], which was able to recognize other three classes: imagination of left-hand movements, imagination of right-hand movements, and generation of words starting with the same random letter. This BCI was based on ERD/ERS patterns from the μ rhythms. Three different classifiers were used in this BCI: LDA (linear discriminant analysis), K-NN (k-nearest neighbors), and ANN (artificial neural network).

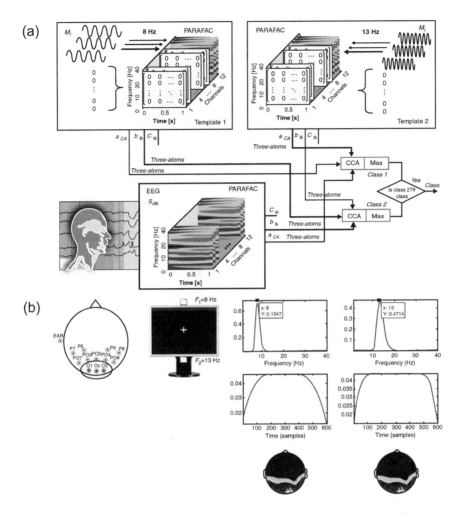

FIGURE 3.13 (a) General schematic of PARAFAC. b) Left: template dimensions (channelsXfrequencyXtime) for 8 Hz. Right: for 13 Hz.

The tasks were processed in the primary motor cortex and in the Broca's area (Figure 3.14). During the imagination of right-hand movements, an ERD appeared in the sensorimotor cortex of the left hemisphere, which could be measured with an electrode located at C3. On the other hand, during the imagination of left-hand movements, an ERD appears in the sensorimotor cortex of the right hemisphere, which can be measured with an electrode located at C4. And during the word generation, an ERD appeared in Broca's area, which could be measured with an electrode at F7. Three success rates for the three classifiers LDA, K-NN, and ANN were 94.3%, 90.3%, and 94.6%, respectively.

Another study [16] analyzed motor imagery through EEG acquisition on F3, Fz, F4, C3, C4, P3, Pz, and P4, at the frequency range of 8 to 32 Hz, using an ANN based on self-organizing maps (SOM). Three classes were considered: right-hand movements,

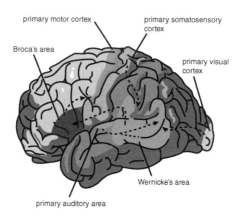

primary motor cortex

Broca's area

primary somatosensory cortex

primary visual cortex

Wernicke's area

primary auditory area

FIGURE 3.14 Location of the primary motor cortex and the Broca's area.

left-hand movements, and remembering of words or imagination of non-motor tasks. The success rate for this classifier reached 72.9%, with an ITR of 58.5 bits/min (Table 3.1).

Another BCI based on motor imagery, using STFT and SVM for feature extraction and classification, respectively, of two (left- and right-feet movements) and three (left- and right-feet movements, and rest) classes applied to filtered signals from 8 to 13 Hz (C3, C4, and Cz), was tested on four people with disabilities (quadriplegia, muscle dystrophy, amyotrophic lateral sclerosis, and cerebral palsy), reaching a success rate of 100% for two classes and 73.4% for three classes (Table 3.2) [17].

Currently, new methods are being applied in our laboratory to improve the success rate of our BCIs for motor imagery detection. For example, STFT, sparseness constraints, and total power in time-frequency representation (to locate the subject-specific bands with the highest power), in addition to Riemannian geometry (to extract spatial features) and a fast version of neighborhood component analysis (to increase the class separability) [18] are used. These methods are being applied to detect the motor imagery to move a robotic exoskeleton [19] (Figure 3.15) or a robotic monocycle (Figure 3.16), trying to provide paralyzed people with some motor recovery, as spinal cord injury (SCI) or poststroke patients may recover some leg movements due to traces of axons between the brain and

TABLE 3.1

Confusion Matrix for the BCI Based on MAPs

10×10	Words	Left Hand	Right Hand
Words	65.93	20.62	13.75
Left hand	9.04	72.89	18.07
Right hand	33.13	25.77	41.10

TABLE 3.2

Results for the BCI Based on STFT+SVM Applied to Four Volunteers with Disabilities

Volunteer	2 Classes (Left/Right)			3 Classes (Left/Right/Rest)		
	ACC	Left Hand	Right Hand	ACC	Left Hand	Right Hand
V1	80.0	80.0	80.0	60.0	60.0	73.4
V2	90.0	80.0	100.0	73.4	60.0	80.0
V3	90.0	80.0	100.0	73.4	60.0	60.0
V4	90.0	100.0	80.0	66.7	60.0	80.0

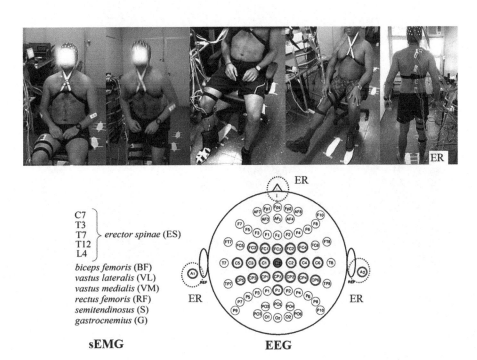

FIGURE 3.15 BCI based on motor imagery to command a robotic exoskeleton.

limbs. Ten subjects, using only 13 EEG channels and only 6 features, have used this BCI in real time, which have obtained 97% of success rate in detecting the motor imagery [22]. This BCI is being used together with an immersive virtual reality system with avatar, as motor imagery can generate more neuroplasticity when combined with an immersive virtual environment and the robotic devices (Figure 3.17) [20].

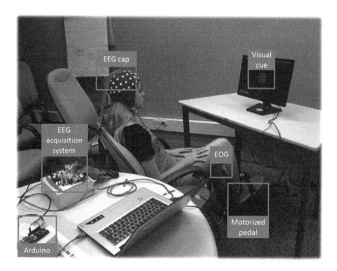

FIGURE 3.16   BCI based on motor imagery to move a robotic monocycle.

FIGURE 3.17   BCI based on motor imagery of pedaling together with an immersive virtual environment.

REFERENCES

1. Ferreira, A; Celeste, W C; Cheein, F A; Bastos-Filho, T F; Sarcinelli-Filho, M; Carelli, R, Human-machine interfaces based on EMG and EEG applied to robotic systems. *Journal of Neuroengineering and Rehabilitation*, 5:1–15, 2008.
2. Bissoli, A; Sime, M; Bastos Filho, T F, Using sEMG, EOG and VOG to control an intelligent environment. *IFACPapersOnLine*, 49:210–215, 2016.
3. Berthil, B L, Development and evaluation of serious games as assistive technology through affordable-access multi devices. PhD Thesis, Universidade Federal do Espírito Santo, Brazil, 2019.
4. Muller, S M T; Bastos-Filho, T F; Sarcinelli-Filho, M; Celeste, W C, Brain-computer interface based on visual evoked potentials to command autonomous robotic wheelchair. *Journal of Medical and Biological Engineering*, 30:407–416, 2010.

5. Diez, P; Muller, S M T; Mut, V; Garci, E; Avila, E; BastosFilho, T F; Sarcinelli-Filho, M, Commanding a robotic wheelchair with a high-frequency steady-state visual evoked potential based brain-computer interface. *Medical Engineering and Physics*, 35:1155–1164, 2013.

6. Garcia, J F C; Muller, S M T; Caicedo, E; Bastos Filho, T F; Souza, A, Non-fatigating brain computer interface based on SSVEP and ERD to command an autonomous car. *Advances in Data Science and Adaptive Analysis*, 1:1–11, 2018.

7. Floriano, A; Bastos Filho, T F, Design of software for visual stimulation of an independent SSVEP-based BCI. *2nd International Workshop on Assistive Technology*, 2019.

8. Foriano, A; Garcia, J F C; Longo, B B; Bastos-Filho, T F, Proposal of a telepresence robot using brain computer interface for people with motor disabilities. *Brazilian Congress on Biomedical Engineering*, 1:2240–2243, 2014.

9. Tello, R; Muller, S M T; Hasan, M; Ferreira, A; Krishnan, S; Bastos-Filho, T F, An independent-BCI based on SSVEP using figure-ground perception (FGP). *Biomedical Signal Processing and Control*, 26:69–79, 2016.

10. Bastos Filho, T F; Floriano, A; Couto, E; Tello, R, Towards a system to command a robotic wheelchair based on independent SSVEP-BCI. *Smart Wheelchairs and Brain-Computer Interfaces*, 1:369–379, 2018.

11. Tello, R; Pant, J; Muller, S M T; Krishnan, S; Bastos Filho, T F, An evaluation of performance for an independent SSVEPBCI based on compressive sensing system. In *World Conference on Medical Physics & Biomedical Engineering*, Springer, Cham., 2015; 982–985.

12. Tello, R M G, A novel approach of independent brain-computer interface based on SSVEP. PhD Thesis, Universidade Federal do Espírito Santo, 2016.

13. BCI Competitions. Available in: http://www.bbci.de/competition/. Accessed on: July 1, 2019.

14. Benevides, A B; Bastos Filho, T F; Sarcinelli-Filho, M, Pseudo-online classification of three mental tasks using KL Divergence. *Journal of Medical and Biological Engineering*, 6:411–416, 2012.

15. Benevides, A B, Proposal of a brain-computer interface architecture based on motor mental tasks and music imagery. PhD Thesis, Universidade Federal do Espírito Santo, 2013.

16. Bueno, L; Bastos Filho, T F, A self-organizing maps classifier structure for brain computer interfaces. *Research on Biomedical Engineering*, 1:1–9, 2015.

17. Ferreira, A et al., Evaluation of PSD components and AAR parameters as input features for a SVM classifier applied to a robotic wheelchair. *International Joint Conference on Biomedical Engineering Systems and Technologies*, 7:12, 2009.

18. Delisle-Rodriguez, D; Cardoso, V; Gurve, D; Loterio, F; Romero-Laiseca, A; Krishnan, S; Bastos Filho, T F, System based on subject-specific bands to recognize pedaling motor imagery: Towards a BCI for lower-limb rehabilitation. *Journal of Neural Engineering*, 1:1–29, 2019.

19. Villa-Parra, AC Admittance Control of a Robotic Knee Orthosis Based on Motion Intention Through sEMG of Trunk Muscles. PhD Thesis, Universidade Federal do Espírito Santo, 2017.

20. Pomer-Escher, A; Loterio, F; Longo, B; Valadao, C; Bastos Filho, T F, Low-cost neurorehabilitation platform for physically disabled subjects using virtual reality and a portable exercise bike. Transactions on Neural Systems & Rehabilitation Engineering, 2019.

21. Rivera, H; Hernandez, K; Longo, B; Bastos Filho, T F, Evaluation of Task Workload and Intrinsic Motivation in a Virtual Reality Simulator of Electric-Powered Wheelchairs. *International Workshop on Future Trends in Assistive Technology 9th International Conference on Current and Future Trends of Information and Communication Technologies in Healthcare (ICTH 2019)*, 2019.

22. Gurve, D; Delisle, D; Romero-Laiseca, A; Cardoso, V; Loterio, F; Bastos Filho, T F; Krishnan, S, Subject-specific EEG channel selection using non-negative matrix factorization for lower-limb motor imagery recognition. *Journal of Neural Engineering*, 17:2, 2020.

4 Future of Non-Invasive BCIs

Teodiano Freire Bastos-Filho

CONTENT

As this book discusses the fundamentals of EEG, describing the human brain, anatomically and physiologically, with the objective of showing some of the patterns of EEG signals used to control brain–computer interfaces (BCIs), as well as a number of very innovative and recent BCIs applied to a robotic wheelchair, an augmentative and alternative communication (AAC) system, a telepresence robot, an autonomous car, several home appliances, a robotic monocycle, a robotic exoskeleton, and an avatar in an immersive virtual environment – which can be also extended to games, one can perhaps foresee the future of non-invasive BCIs.

Actually, very little objective discussion has been carried out in the literature about the general usability of these BCIs, such as the comfort necessary to wear the EEG acquisition system for long term, their efficiency, and possible health problems provoked by their use.

First of all, it is important to highlight that currently BCIs have several limitations; for instance, they rarely are 100% accurate (which also implies in low information transfer rate or ITR), which may depend on the classifier (different classifiers are used in several BCIs, and to decide which one is the best one, it would be necessary to test them within the same context, i.e., with the same users and protocol, using the same methods for both feature extraction and dimensionality reduction or feature selection – using, for instance, EEG data from international databases (for instance, from BCI Competitions [1]). Currently, adaptive classifiers are considered the best ones [2,3]). Also, several BCIs use supervised methods instead of non-supervised ones, such as, for instance, those described in Ref. [4–6], whose performance can be affected by patterns selected in the learning stage, as some users are unable to learn and execute the protocol correctly. For instance, in Ref. [7], around 17% of the participants were unable to learn to perform the BCI protocol. It is a fact that BCIs do not work for everybody, as some users never reach accuracy above 70% (named BCI illiterates) [8–12]. In the literature, there is a common understanding among researchers that BCIs with accuracy below 70% do not allow useful operation [13–18]. In addition, BCIs normally require a lot of training and, for long-term use, re-calibration is always needed.

With respect to the EEG signal acquisition using electrodes with gel (considered the gold standard, and for this reason, it is the technique applied in clinics), the preparation time is a laborious process that begins with the localization of sites for the electrode montage. Then, to reduce the contact impedance between the electrode and the skin to acceptable values (typically below 5 kΩ), these sites are rubbed with an abrasive paste that removes part of the outer skin (stratum corneum), which is the main contributor to the high skin impedance. Following, typical Ag/AgCl electrodes are gel-soaked, in order to facilitate the transduction of the ionic currents (which freely move through brain tissues and cerebrospinal fluid) into electric currents [19]. Then, the electrode–skin contact impedance must be measured to guarantee a low value and start the signal acquisition. All this preparation can take a long time. On the other hand, the use of abrasive paste and electrodes with electrolyte gel makes the hair and scalp dirty, which is very inconvenient. For SSVEP-based BCIs, a solution to diminish that inconvenient is not using a cap with electrodes, but only 1–2 electrodes on the scalp, such as done by [20]; however, in that case, the BCI accuracy drops significantly. Anyways, after approximately 5 h, the contact impedance between the electrode and skin starts to deteriorate, and it is necessary to re-apply the gel to get high-quality EEG signals [21]. Another solution is placing electrodes below the hairline areas, as described in Ref. [22,23]; however, the accuracy of these SSVEP-based BCIs drops significantly in this case as well.

Because of the laborious process for the montage of electrodes with gel and the inconvenience of making the hair and scalp dirty, Emotiv Systems launched the EPOC neuroheadset in 2009, which uses wet electrodes. Actually, these electrodes are connected to saline-soaked foam pads, which are capable of establishing a connection through most types of hair. This neuroheadset is advantageous to users, as the only residue left in their hair is saline (whereas gel and paste are sticky products that make the hair and scalp dirty). In fact, some authors consider the EPOC neuroheadset as the best performance/price/design ratio in the consumer segment, with a cost below a thousand of US dollars [24]. However, a limitation for this neuroheadset is that, for long-term experiments (>15–20 min), the EEG signal quality degrades as the contacts dry out [25,26], and some users complain that it hurts the head after an hour of use [27].

Recently, EEG dry electrodes have arisen in the market, which makes unnecessary the use of gel, paste, or saline, however, in a very expensive cost: a cap with 16 dry electrodes and accessories for EEG acquisition can cost some thousands of US dollars. In addition, some of these dry electrodes can cause pain, due to the high level of contact pressure on the head produced especially by using comb electrodes with pointy tips. Other kinds of EEG headsets also cause pain, although in lower level [28,29].

Thus, although BCIs can be successful for many applications, such as the ones shown in this book, they are still not completely comfortable for users. As a portable unobtrusive technique to acquire EEG signals is still unavailable and the electrodes with gel are the gold standard to acquire high-quality EEG signals (although this technique needs long preparation time and makes the hair and scalp dirty, and the gel dries after some hours), I believe that the challenge to be faced by future BCIs is to deploy low-cost dry electrodes with the ability to acquire EEG signals with the same level of quality of electrodes with gel. On the other hand, the headset with dry electrodes should have contact adjustments to apply just the right amount of pressure (even for different head sizes), in order to avoid pain due to over-tightening.

As reported in Chapter 1, BCIs based on SSVEP, depending on their frequency range, can cause epileptic seizures, migraine headaches, and visual fatigue. Epileptic seizures can be avoided in many applications by not using the frequency range from 15 to 25 Hz; migraine headaches and visual fatigue can be avoided using the strategy detailed in Section 3.2.3, which suggests eye closing (which affects the ERD/ERS patterns) to turn off the flickering.

Finally, regarding the fact that no BCI reaches 100% accuracy for everybody in all moments, in applications in assistive technology, a BCI with at least 70% accuracy could be quite useful, as a 30% of error may be acceptable for people with severe disabilities, who may have no other options to interact with the people and the world around.

REFERENCES

1. BCI Competitions, Is the Emotiv EPOC signal quality good enough for research? Lab Talk. Available in: http://www.bbci.de/competition/. Accessed on: January 9, 2020.
2. Delisle-Rodriguez, D; Villa-Parra, A C; Bastos-Filho, T F; Lopez-Delis, A; Frizera-Neto, A; Krishnan, S; Rocon, E, Adaptive spatial filter based on similarity indices to preserve the neural information on EEG signals during on-line processing. *Sensors*, 17:2725, 2017.
3. Lotte, F; Bougrain, L; Cichocki, A; Clerc, M; Congedo, M; Rakotomamonjy, A; Yger, F, A review of classification algorithms for EEG-based brain-computer interfaces: a 10 year update. *J. Neural. Eng.*, 15(3): 1005, 2018.
4. Jiang, N; Gizzi, L; Mrachacz-Kersting, N; Dremstrump, K; Farina, D, A brain-computer interface for single-trial detection of gait initiation from movement related cortical potentials. *Clin. Neurophys.*, 126:154–159, 2015.
5. Xu, R; Jiang, N; Lin, Ch; Mrachacz-Kersting, N; Dremstrump, K; Farina, D, Enhanced low-latency detection of motor intention from EEG for closed-loop brain-computer interface application. *IEEE Trans. Biomed. Eng.*, 61(2):288–296, 2014.
6. Hashimoto, Y; Ushiba, J, EEG-based classification of imaginary left and right foot movements using beta rebound. *Clin. Neurophys.*, 124:2153–2160, 2013.
7. Jeunet, C; Jahanpour, E; Lotte, F, Why standard braincomputer interface (BCI) training protocols should be changed: an experimental study. *J. Neural. Eng.*, 13(3): 036024, 2016.
8. Allison, B Z; Neuper, C, Could anyone use a BCI?. *Applying our Minds to Human-Computer Interaction*, eds. D. S. Tan and A. Nijholt (London: Springer Verlag), 35–54, 2010.
9. Allison, B Z; Valbuena, D; Lueth, T; Teymourian, A; Volosyak, I; Graser, A, BCI demographics: how many (and what kinds of) people can use an SSVEP BCI?. *IEEE Trans. Neural. Syst. Rehabil. Eng.*, 18:107–116, 2010.
10. Allison, B Z; Brunner C; Kaiser V; Muller-Putz G R; Neuper C; Pfurtscheller G, Toward a hybrid brain-computer interface based on imagined movement and visual attention. *J. Neural. Eng.*, 7(2): 26007, 2010.
11. Brunner, C; Allison, Krusienski D J; Kaiser V; Muller-Putz G R; Pfurtscheller G; Neuper C, Improved signal processing approaches in an offline simulation of a hybrid brain-computer interface. *J. Neurosci. Methods*, 188(1): 165–173, 2010.
12. Rimbert, S; Gayraud, N; Bougrain, L; Clerc, M; Fleck, S, Can a subjective questionnaire be used as brain-computer interface performance predictor?. *Front. Hum. Neurosci.*, 12:529, 2019.

13. Holz, E M; Botrel, L; Kubler, A, Bridging gaps: long-term independent BCI home-use by a locked-in end-user. N.-L. Millan, N. Guechoul, R. Leeb, & J. del R. Millan (Eds.), *Proceedings of TOBI Workshop IV*, Sion, Switzerland, 35–36, 2013.

14. Volosyak, I; Valbuena, D; Luth, T; Malechka, T; Graser, A, BCI demographics II: how many (and what kinds of) people can use a high-frequency SSVEP BCI? *IEEE Trans. Neural. Syst. Rehabil. Eng.*, 19(3):232–239, 2011.

15. Dornhege, G; Millan, J R; Hinterberger, T; McFarland D; Muller, K R, *Towards Brain-Computer Interfacing*. Cambridge: The MIT Press, 2007.

16. Blankertz, B; Sannelli, C; Halder, S; Hammer, E M; Kubler, A; Muller K R; Curio, G; Dickhaus, T, Neurophysiological predictor of SMR-based BCI performance. *Neuroimage*, 51(4):1303–1309, 2010.

17. Fernandez-Vargas, J; Pfaff, H U; Rodriguez, F B; Varona, P, Assisted closed-loop optimization of SSVEP-BCI efficiency. *Front. Neural. Circuits*, 25(7):27, 2013.

18. Kubler, A; Neumann, N; Wilhelm, B; Hinterberger, T; Birbaumer, N, Predictability of brain-computer communication. *J. Psychophys.*, 18(2–3):121, 2004.

19. Lopez-Gordo, M A; Sanchez-Morillo, D; Pelayo-Valle, F, Dry EEG electrodes. *Sensors (Basel)*, 14(7):12847–12870, 2014.

20. Muller, S M T; Bastos-Filho, T F; Sarcinelli-Filho, M, Monopolar and bipolar electrode settings in SSVEP-based brain-computer interface. *J. Med. Biol. Eng.*, 1:1–12, 2015.

21. Lin, C T; Liao, L D; Liu, Y H; Wang, I J; Lin, B S; Chang, J Y, Novel dry polymer foam electrodes for long-term EEG measurement. *IEEE Trans. Biomed. Eng.*, 58(5):1200–1207, 2011.

22. Floriano, A; Delisle-Rodriguez, D; Diez, P; Bastos-Filho, T, Assessment of high-frequency steady-state visual evoked potentials from below-the-hairline areas for a brain-computer interface based on depth-of-field. *Comp. Methods Prog. Biomed.*, 1:105271, 2019.

23. Floriano, A; Diez, P; Bastos-Filho, T, A study of SSVEP from below-the-hairline areas in low-, medium-, and highfrequency ranges. *Res. Biomed. Eng.*, 35:7176, 2019.

24. MAMEM Consortium, Initial integration and optimization of multi-modal sensors. Multimedia Authoring and Management using your Eyes and Mind. Available in: https://www.mamem.eu/wp-content/uploads/2016/12/D2.2_Initial_%CE%99ntegration_Optimization_Multi-modal_Sensors_Final.pdf

25. Sapiens Lab, Is the Emotiv EPOC signal quality good enough for research? Lab Talk. Available in: sapienlabs.org/emotiv-epoc-signal-quality-good-enough-research/ Accessed on: January 6, 2020.

26. Taherian, S; Selitskiy, D; Pau, J; Claire-Davies, T, Are we there yet? Evaluating commercial grade brain-computer interface for control of computer applications by individuals with cerebral palsy. *Disabil. Rehabil. Assist. Technol.* 12:165–174, 2017.

27. Mayaud, L; Congedo, M; Van Laghenhove, A; Orlikowski, D; Figère, M; Azabou, E; Cheliout-Heraut, F, A comparison of recording modalities of P300 event-related potentials (ERP) for brain-computer interface (BCI) paradigm. *Neurophysiol Clin.* 43(4):217–227, 2013.

28. Fiedler, P; Muhle, R; Griebel, S; Pedrosa, P; Fonseca, C; Vaz, F; Zanow, F; Haueisen, J, Contact pressure and flexibility of multipin dry EEG electrodes. *IEEE Trans. Neural. Syst. Rehabil. Eng.*, 26(4):750–757, 2018.

29. Verwulgen, S; Lacko, D; Justine, H; Kustermans, S; Moons, S; Thys, F; Zelck, S; Vaes, K; Huysmans, T; Vleugels, J; Truijen, S, Determining comfortable pressure ranges for wearable EEG headsets. *International Conference on Applied Human Factors and Ergonomics (AHFE): Advances in Human Factors in Wearable Technologies and Game Design*, 11–19, 2018.

Index